Gerald Leach

# THE BIOCRATS

Implications of Medical Progress

Revised Edition

Penguin Books

Penguin Books Ltd, Harmondsworth,
Middlesex, England
Penguin Books Australia Ltd, Ringwood,
Victoria, Australia

First published by Jonathan Cape 1970
Revised edition published in Pelican Books 1972
Copyright © Gerald Leach, 1970, 1972

Made and printed in Great Britain by
Richard Clay (The Chaucer Press) Ltd
Bungay, Suffolk
Set in Monotype Times

PELICAN BOOKS
The Biocrats

Gerald Leach is science correspondent of the *Observer*.
Born in Ceylon in 1933, he read natural science at
Cambridge. His early career as a science writer included a
period as assistant science correspondent on the then
*Manchester Guardian*. Between 1959 and 1961 he worked
in television – writing and presenting the weekly science
programme 'It Can Happen Tomorrow', co-producing
several science documentaries, including nine editions of
'Eye on Research', and working for a year with Dr
Bronowski on the eleven-programme epic 'Insight'. From
1961 to 1963 he was editor of the scientific monthly,
*Discovery*, at the time the only semi-popular magazine of its
kind in Britain. From then until he took up his position
with the *Observer* in 1969, Gerald Leach was a freelance
science writer, spending two years as science correspondent
of the *New Statesman*, working on many television and
radio science programmes, and conceiving *The Biocrats*.

To Penny, who had a hard time of it

*A note on the cover picture*

This shows a seven-week-old living human foetus. It was obtained by Caesarian termination of pregnancy, and the foetal heart was kept beating for several hours by immersing the foetus with its surrounding amniotic fluid and transparent amniotic sac in a bath of oxygenated saline solution.

Technique and photograph by Dr Landrum B. Shettles, Columbia-Presbyterian Medical Center, New York City.

# CONTENTS

Acknowledgements  9
Introduction  11

1. Birth-control  18
2. Population Control  52
   Interlude: On Keeping Calm  77
3. Test-Tube Reproduction  80
4. Breeding For Quality  117
5. Breeding Out Faults  126
6. Foetal Medicine  161
7. Birth Defects  187
8. Brains  215
9. Life on the Machine  246
10. Transplants  284
11. Priorities  326

Index  357

# ACKNOWLEDGEMENTS

This book is mainly intended for a general audience, but has at least half an eye on the experts too. The first kind of reader is usually maddened by copious references for every fact, figure and quotation; the second is maddened if they are missing. So I have compromised by giving references only where the facts seem to me surprising or little known, or a source is used heavily.

I have consulted many hundreds of papers and books while writing this volume, some of them brilliant and stimulating, and I would like to express my debt and thanks to the legion of unnamed authorities who thus helped me so much. Especially, I would like to thank most warmly the scores of scientists and doctors in the United States and Britain who gave me their time to talk or show me what they were doing.

Last but not least I owe a particular debt to Mr John Wolfers, Mr Frank Taylor, Mr Dieter Pevsner, my wife and her family for starting me off and for their patience and encouragement until I finished; and to Miss Caroline Durston and Mrs Ruth Stungo for help with research. May I hasten to add that none of these helpers – named or unnamed – necessarily agrees with the views I have expressed in the book.

# INTRODUCTION

The human race is moving towards the future through an endless branching maze of incredible complexity. Every day, at each point in the maze marked 'now', we see a thousand paths spreading out before us, each leading to a slightly different possible future – and another point where a thousand paths diverge. And at each point we have to choose which few paths to take, and which to leave alone. By these choices – some of them trivial in the long run, but some of them very significant – we arrive in one new part of the maze, or another. We choose our own future, for better or for worse.

It is not an easy game, and it is made no easier by science and technology. At an ever-increasing rate, science and technology are gaining far-reaching powers to shape the world, and ourselves with it. They, and their implications, are also becoming more difficult to understand. Just as they are complicating the maze by creating more and more possible paths and creating them more frequently, so it is becoming harder for most people to help to choose the right ones. As the maze grows more intricate and the decision points rush at us faster, the walls of the maze are becoming opaque; and we face a growing danger of being led up the wrong paths by the ignorant or by the forceful but blinkered army of technocrats.

This book is an attempt to help us to steer our way through one of the most complex and important parts of the maze. It is no secret that biology and medicine are rapidly discovering the possibility of very radical alterations to the human body, brain and life processes; and that these new powers closely touch the deepest human and social values. From dealing with sickness and disability, medicine is being pushed by modern biology into a control of the human machinery so intimate and pervasive that it has profound consequences for the basic aspects of life – sex, procreation, birth, the relationship between parents and offspring, the nature of the individual, his role in society and the nature of society itself. Perhaps more than any other science or technology, biology and medicine threaten to make the changes that will alter men and societies most. We must therefore learn to steer our way through the biological maze with all the courage and insight we can muster. If we do not, we may find ourselves in a part of the maze we do not like at all *and* miss the paths that could have taken us instead to a far better place.

So far we have not managed this well. Time and again society has woken up to find itself living with a major biomedical trend that it has never agreed to but which is now very hard to alter. The technocrats of biology and medicine – the biocrats – had taken charge.

The population explosion is the best example: because vastly improved techniques of death-control were used in the developing world before equally effective techniques of birth-control were available and acceptable, the population explosion was inevitable. But we lacked the social mechanisms for predicting this, for debating it and for applying the necessary correctives in time. It has been much the same with many other developments, from transplants to artificial organs, from intensive care for disabled babies to intensive care for the old. And unless we change our ways it looks as though it will be the same for the more recent, less well-known trends mentioned in this book.

Why is this? A short answer is that all the key groups of people who should be contributing to the debate and decision-making cannot do so effectively because they are walled off by their own expertise and attitudes, and do not listen to those outside their narrow confines.

There are the biologists, the generators of the biomedical revolution. They contribute little here because, like other scientists, they have no set of principles concerned with the social effects of their work. They have many important ethical attitudes – the most important of which is that they tell each other the truth about what they are doing – but their attitudes are all concerned with protecting the profession of science. There is no provision innate in science to ensure that it concerns itself with the needs of society. Indeed, most scientists insist that they must be free to make any discovery they can, whatever its social implications. As far as 'pure' science is concerned – science restricted to experimental results on paper – this does not matter in the narrow view and the short run. As Bertrand Russell once said, 'equations do not explode.' But it does matter overall and in the long run, because it loosely affects what kinds of things scientists set out to discover and it directly affects what ideas scientists produce that *can* be applied. Though there are notable exceptions, it also produces scientists who cannot be bothered to tell the public of the implications of their work – after thinking hard about them (a necessary step that is not always taken).

As discoveries *are* applied, as they move from the laboratory into the hospital or surgery, we run into the notorious closed-shop, shut-mouth mystique of medicine. This tradition of silence was crucially important at a time when all the doctor could do to cure the sick was to sit at the bedside, radiate a sense of knowing calm and confidence and let nature take its course. For if it broke down, and the patient was allowed to look in the doctor's black bag, he would find it empty – and faith would be shattered. Today, when the bag is overflowing with therapies that the patient and society may or may not want, it is a dangerous anachronism.

The same applies to the other more important medical attitudes, which we loosely group together under the heading of 'medical ethics'. Traditional medical ethics are mainly concerned with putting the welfare of the individual patient first, to the exclusion – if necessary – of other interests, and with doing all one can to save life. These were and are important principles. But as the escalating power of medical technology produces greater consequences for individuals and society, they are rapidly becoming hopelessly inadequate on their own. Most people, including many doctors, realize that these principles have to be broadened to take into account not only the patient's wishes seen against his home and family background (rather than merely his medical condition), but also the interests of society. While it was once a triumph to be able to save life by putting a patient in an iron lung for twenty years, it is not self-evidently so now. The patient may rather be dead than alive like that; and society might want to spend the money in other ways.

This broadening of principles can create agonizing conflicts of conscience for the doctor and difficult decisions for the patient. But the only way of relieving these pressures is to try to form a social consensus by *talking* about them, so that new principles that most people will accept naturally may emerge. If there are *no* iron lungs because it is agreed that there should not be, then dying because one is totally paralysed becomes easier to bear. In a similar way, taking the life of a foetus is now easier to bear – for doctors, parents and society – because we have begun a debate on abortion that is changing our principles in this field. Yet this is an exception: with most areas of medical practice the medical profession still prevents *informed* public discussion by its determination to hold on to its right to make the decisions.

But assuming that informed public discussion might be possible we face another obstacle, which is that the principles which have traditionally formed the basis of such discussion are crumbling. For this reason our former advisers – theologians, philosophers, anthropologists, lawyers and so on – are no longer able to provide the foundations on which a discussion of the ethical problems of medicine should be based – even if they were well informed on medical advances.

So we are forced to reconsider beliefs which up to now have been thought of as constants, and decide which we wish to preserve and which to reject. It seems to me, for instance, that the family is a constant which we ought to preserve. This often reviled unit is universal in some form or other throughout the human race and has never been abandoned successfully for long. We ought to think carefully about developments that threaten it. On the other hand, the concept of genetic lineage – the notion that only the couple who conceive a child can be his 'true' parents – is an age-old constant that we might abandon without much trouble. Other examples, of both kinds, will recur throughout this book.

For the moment, the important thing about these problems concerning belief is that, with one crucial proviso, they are not issues which biologists or doctors are either more or less qualified to discuss than any other comparably educated layman. The tension between them and biomedical progress *can* therefore be discussed – and must be discussed if we are to find the best futures in the maze – by everyone who reckons himself a member of the community, backed by whatever help the social thinkers can give him. We must not let the biocrats, whether biologists or doctors, make all the running.

The one proviso is that we all have to be properly informed about the technical side too – the biology and medicine. The fact is that we are not. Most people are unable to enter most areas of the biomedical debate because they are technically ill-informed or misinformed. It is not that they do not know the details or the jargon: they rarely matter. It is that most people have not even the equipment to discriminate between the technical probabilities, improbabilities and impossibilities of the foreseeable future. Neither, very often, can they distinguish between techniques that people will want to take up and those they will want to leave alone. How else can one explain all those earnest television discussions of brain transplants, or articles hinting at the imminent possibility of rearing babies entirely in test-tubes à la *Brave New World*? We might just as well spend our time seriously discussing the ethical implications of an invasion by Martians.

The blame here must fall squarely on the mass media, the source of most people's knowledge for the biomedical debate. Medicine is always news, but how much more newsworthy when one can shout of a new miracle cure (and not cite all the disadvantages) or cry 'scandal' (and not bother to discover the real reasons why the situation is as it is). When this over-dramatization is accompanied by instant predictions of what the biologists are cooking up for the future – visions ranging from the apocalyptic to the euphoric depending on the writer and his mood – no wonder our expectations are totally confused.

For these reasons quite a large proportion of this book is taken up with surveying the facts and the practical implications of different areas of biochemical activity as fully and objectively as possible. Only when we have this basic information is there any point in discussing the ethical, human and social implications. Where these 'softer' subjects are discussed, again I have attempted to be objective by putting forward the major arguments for and against.

A second characteristic of this book that needs explaining is that it covers few ventures into the future of more than a decade or so ahead. The technical future is coming at us very fast and is almost completely obscure beyond a ten-year or at most fifteen-year time-span.

This is because new discoveries take about this period to reach application. If we look forward within this limit we can get a rough idea of what may be going to happen based on what we know now; though even within this span we may miss crucial details of the form these applications will take, or the scale on which they will be taken up. For example, with the ability to pre-select the sex of our children the form of the technique – perhaps swallowing a pill, perhaps artificial insemination – makes all the difference to the scale on which it will be taken up, and hence the scale of the problems it raises.

If we try to look further ahead – say to the year 2000 – we can not only guarantee to get these aspects wrong; we will almost certainly fail to predict most of the most influential discoveries themselves. In 1937 a high-level group of American scientists attempted to predict the important technological developments that would be made by 1967.* Among other things, they missed out computers, radar, most contemporary electronic devices, effective antibiotics and the jet engine, and thought that aircraft would get safer and more comfortable but not appreciably faster. There is no indication that we can do any better today. So, with the possible exception of human genetic engineering, all the developments

* *Technological Trends and National Policy. Including the Social Implications of New Inventions* (June 1937). *Report of the Subcommittee on Technology to the U.S. National Resources Committee* (U.S. Government Printing Office, Washington, 1937).

discussed here already have their foundations laid in present-day biology and medicine.

These foundations are laid, of course, in the advanced nations. They will be first applied in the advanced nations, and it is in them that they must first be debated and controlled. Indeed, it is only in the advanced nations that we have the time and energy to worry about them. When you have more than thirty thousand people for every doctor (as in most of Africa and Indonesia) you are less worried about the effect of eugenics or personality-affecting drugs or transplants on the values of society than you are about the effect of the lack of medicine on your dying child.

It can be argued that until no one is dying for lack of food or simple medicine, until twenty million people are no longer blind from the trachoma virus for lack of simple hygiene or an anti-biotic swab, until millions of city dwellers no longer have to wade through their own disease-laden excrement when it rains, debating the issues raised in this book is an intellectual luxury. If one looks at the ethics of medicine on a global scale, the problems of the advanced countries can seem provincial and puny. But however unjust it is, Western technology does race ahead of global needs, and the problems which it has brought and will continue to bring in these areas are real. To ignore them because one feels guiltily that they are not morally as important as others is to put one's head in the sand. We have been doing that for too long.

# 1. BIRTH-CONTROL

The most striking thing about birth-control is the yawning gap between the 'cans' and the 'cannots'. While a small fraction of the human race can now prevent sex having disastrous consequences, the majority cannot. Because of a staggering hotchpotch of taboos, moral confusions, social restrictions and ignorance, they still live in a miserable prison of sexual and reproductive failure.

While this gap exists I have no hesitation in starting this book with what may seem like a well-worn topic. For one thing, the birth-control gap still matters enormously in terms of the happiness of people – and of populations. There are few medical topics more significant.

But, equally important, birth-control is a perfect illustration of the book's main theme that we cannot hope to reap the benefits of the biomedical revolution unless we blast most of our most deep-seated values about 'life' into the open, examine them carefully, and ruthlessly discard those that are obsolete. With birth-control we have been extremely prissy and evasive about doing this, and it is time we did better. Unless we do, we will have a very rough time with some of the more novel techniques on the biomedical horizon. Also, because so many of these techniques are to do with reproduction (Chapters 2 to 7) the relatively familiar territory of birth-control is an excellent stamping ground for sorting out which values are important and which are disposable relics.

Before looking at these values, let us look at some facts about birth-control. The most significant is that the birth-control gap is also the gap between those who use 'open' methods and those who use 'closed' ones.

'Open' methods are open for all to use. They were nearly all invented by the time of the Stone Age, and include child-killing, self-induced or 'back-street' abortions, withdrawal or coitus interruptus, and the male sheath or condom.

Taken together they can be surprisingly effective in preventing unwanted children. A triple-layered defence system works best, with baby-killing backing the illegal abortionists, who in turn back the front-line troops – the millions of couples who so much do not want to conceive that they are willing to savage their sex lives in the attempt. This is how Western Europe brought its birth-rate crashing down in the nineteenth century, and it is how at least 90 per cent of all men and women today try to make sex harmless.

But they are hardly effective in enhancing human happiness. Consider, for instance, the situation in the Catholic-dominated South American country, Colombia. Reporting to the World Health Organization in 1967, the Colombian Minister of Health claimed that most women of thirty-five to forty had six or seven children, more than half of them unplanned or unwanted. To stop more coming they try to use primitive contraception, but it usually fails. So 'the women turn from their men and frigidity is increasing. We find that 50 per cent of wives have no orgasms any more and reject their husbands. So the husbands look for other women and the family breaks down.'

When this first line of defence has failed, the women may try abortion. Medical abortions are almost entirely illegal, so the women have to risk their lives with back-street operators. As a result, abortion is the largest single cause of death among women under forty (9 per cent of all deaths) and 44 per cent of all women in hospital are there because of botched abortions.

If this second line of defence fails, there is always infanticide. It is now known that a good deal of childhood malnutrition – and death – in Colombia is deliberate: 'Many women are underfeeding their children, letting them get sick, throwing them outside the house, and then taking them to the doctor knowing it is too late.'

Turning to the advanced countries the picture is far better. Yet despite all the talk about 'permissive societies based on the pill' there is an enormous way to go before one can look at it without shame.

Consider the situation in Britain, which since the 1967 Family Planning Act and the 1967 Abortion Act has some of the most progressive birth-control legislation in the world.

Infanticide is now rare, yet two hundred dead babies still turn up in London dustbins each year, and one still comes across stories like this in newspapers: 'A 15-year-old girl had a baby in her bedroom, helped by her 13-year-old sister, and without the parents knowing, and then went back to work normally the next morning after strangling the child . . .' (*Guardian*, 15 March 1969).

Safe abortions under good medical conditions are now allowed on broad 'social' grounds in National Health Service hospitals (free) or in the private sector (at a cost usually between £50 and £150). As Table 1.1 shows, since the Abortion Act came into force in April 1968, this type of abortion has increased dramatically – presumably largely at the expense of illegal, back-street operations. Yet as the high fees and level of business in the private sector shows, the demand for abortion still far outstrips the supply – especially in some regions where the abortion rate is a half or less of the national average. Tens of thousands of women still cannot get an abortion in time, or cannot afford private fees, and so turn to back-street abortionists or have an unwanted baby.

Roughly 200,000 pregnancies a year – or one in six of all pregnancies – are unwanted: these include all abortions, most illegitimate births and a proportion of pre-marital conceptions.

About one in twelve babies is illegitimate, or 67,000 out of roughly one million births. The illegitimacy rate has doubled in the last decade, though since 1969 the rise in illegitimacy has halted and begun to reverse – possibly as a consequence of the Abortion Act.

Nevertheless, tens of thousands of babies are still made legitimate only by the expedient of a 'shot-gun' marriage. Each year more than 90,000 babies – or one quarter of all first-borns – are born within eight months of marriage. Though many of these marriages would have been made anyway, clearly a very large number are due mainly to parental or social pressures – with dire

TABLE 1.1    Abortions in Britain since the Abortion Act

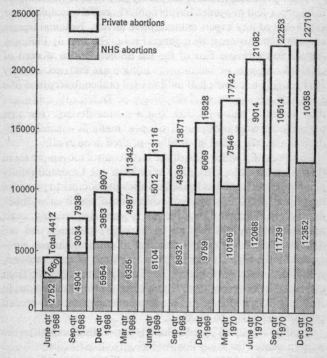

results. One quarter of teenage marriages where the bride was pregnant end in divorce, compared to 9 per cent of all marriages.*

Various estimates suggest that about half of all pregnancies – inside and outside marriage – are basically unplanned, meaning

*Griselda Rowntree to the International Congress on Mental Health, London, 1968, reported in *The Times*, 13 August 1968. Other figures given here come from the Family Planning Association, or the Registrar General's *Statistical Reviews*.

that if they are not actually unwanted they have 'just happened' rather too early. Some surveys show that when women have had several children, more than four out of five of them admit that their last child was either unwanted or unplanned.*

As for contraceptive use, the survey just cited found that less than 20 per cent of women having babies in two city hospitals had ever received any expert contraceptive advice. National figures confirm that this must be a general rule. The Family Planning Association estimate that of the ten million British women of child-bearing age, of whom seven million are married, in early 1971 roughly one and a half million used oral contraceptives (the pill), half a million used a diaphragm or Dutch cap, and some fifty thousand had an I.U.D. (intra-uterine device). The vast majority therefore used primitive 'open' methods – mainly withdrawal, the safe period or sheaths – or used none at all.

For a country which started the birth-control movement more than a century ago, which has more than one thousand family planning clinics, and has legislation allowing doctors to prescribe the pill and diaphragm to anyone over sixteen, that is an astonishing pattern. Though use of the pill is increasing steadily, most married couples still rely on contraceptive methods that are neither efficient nor exactly designed to enhance the sex act – especially the classic twins of 'open' methods of contraception, withdrawal and the safe period. The Family Planning Association estimated in the spring of 1971 that some four million women in Britain still lacked effective, modern contraception and needed professional advice to obtain it.

In other 'advanced', predominantly non-Catholic countries the pattern is much the same. For instance, in mid-1967 Australia and New Zealand had 35 per cent, and the U.S.A. and Canada 25 per cent of married women of child-bearing age on the pill. Most of the remainder used the safe period, withdrawal (less common than in Britain) and the sheath.

The only real contrast is in the Catholic countries of Europe and South America, where the sole permissible birth-control

* E.g. A. C. Fraser and P. C. Watson, 'Family Planning – a myth?', *Family Planning* (October 1968).

method is the safe period.* Despite this, the overwhelmingly important methods are the safe period, withdrawal and abortion (in Italy there are an estimated million abortions a year, or one for each live birth). However, use of the pill is increasing rapidly, if illegally.

In sharp contrast to these overall patterns of 'open' birth-control, with their stench of failure and sexual frustration, there are the 'closed' methods enjoyed – as the figures hint – by the few. These are the effective, sexually silent methods: medical abortion, sterilization, I.U.D.s, the diaphragm and oral contraceptives.

I call them 'closed' because they have one thing in common: they are all medical methods, and before using them one has to confront a doctor backed by all the moral attitudes of his profession and the religious and official institutions of society.

If the birth-control gap is to be narrowed, the great questions are whether these attitudes are valid; if they are not, how can they be removed; and if they cannot be removed at all easily – as I suspect – is there any chance that technology can by-pass the whole tangled mess and produce effective *open* methods of birth-control which anyone can buy over the counter? Now let us divide birth-control into its other two main divisions – contraception and abortion – and look at these questions in turn.

## Contraception

One of the main attitudes blocking the spread of modern contraception is the persistent belief that sex is not for pleasure but for procreation. This, of course, is the basic Catholic position: the sex act is for the transmission of life and to deny this is to turn a woman into a mere 'receptacle for lust'. Therefore celibacy, the safe period† – or sin.

* Except (at the time of writing) for France, which legalized the pill and other mechanical methods for everyone over eighteen in February 1969; and Italy, which abolished laws prohibiting the manufacture, sale and use of all contraceptives in March 1971.

† The safe period is so unsafe that if 100 average women use it for a year about 25 will become pregnant. Women's reproductive cycles vary

But despite *Humanae Vitae* (and probably because of it), a growing number of Catholics now realize what logical and human nonsense this is. They see, like the rest of us, that the main human purpose of sex is not procreation but pleasure (meaning anything from the gratification of lust to the total gift and abandon of mutual love).

In fact for humans, as opposed to animals, pregnancy usually represents a *failure* of the main biological purpose of sex, which is to gratify a couple so that they are more likely to stay together and care for their offspring – offspring which take a uniquely long and helpless time to mature. Why else should the human sex drive, unlike that of any other animal, be geared to act continuously?

This confusion over the purpose of sex is at the root of the second main attitude blocking the spread of good contraception. This is the 'big stick' argument. By refusing to let the unmarried have good contraceptives, the argument goes, one can beat down sexual sin and keep closed the 'floodgate to promiscuity' (a favourite phrase).

Of course, one *can* keep down sexual activity through the fear of pregnancy. It was a major factor a century ago, as can be seen in almost any Victorian novel. It is still a factor today. In the mid-1960s, for example, Michael Schofield studied the sexual behaviour and attitudes of nearly two thousand unmarried teenagers in Britain with one of the best-conducted surveys ever done in this field.* Despite popular myths about teenage sexual

---

so much that many couples have to avoid intercourse for 10 days a month, and some for 15 days, if they are to be completely safe. Since one in six women is so irregular that she cannot use the method at all, in the long run it is self-defeating. If all women used it, as Catholic dogma insists they should, the irregular would be bound to outbreed the regular, so that all women would one day be irregular. There would then be no more 'safe' periods. There is also controversial evidence that because the safe period increases the chance of fertilizing the egg (ovum) when it has begun to deteriorate a few days after ovulation, it increases the chance of producing babies with physical or mental defects.

* M. Schofield, *The Sexual Behaviour of Young People* (Longmans, London, 1965; Pelican, 1968).

performance, he found that two-thirds of the boys and three-quarters of the girls had never had any sexual experience. He also found that fear of pregnancy does rate high as a disincentive.

But his most striking discovery was the extent of ignorance about birth-control. While 80 per cent of both sexes knew that something called birth-control existed, 20 per cent did not. Only 45 per cent of boys and 20 per cent of girls with sexual experience insisted on some precautions – usually a sheath – as a condition of intercourse. The others, the non-users, gave the unavailability of contraception (32 per cent), dislike of the results (21 per cent), or 'don't care' (20 per cent) as their reason.

Surely, what figures like these demand is that those who wish to reduce sexual 'immorality' should ask themselves just what they mean by it. Effective contraception has blown the morality of sex into two pieces. By divorcing the sexual act from its consequences it insists that we now treat the moralities and responsibilities of each quite separately.

There is the morality of sexual intercourse; and the morality of the reproductive consequences of intercourse. We have to ask a new set of questions. Is all extra-marital sex immoral, even though both parties treat each other generously and maturely? Or is it immoral only when one partner selfishly exploits the other for his own sexual gain? And, as regards the consequences, is not the greatest immorality of all wilfully to risk producing an unwanted child (or the often tragic alternative, the shot-gun marriage)?

Until parents, teachers and all the other moral guardians of society become less evasive about questions like these, we shall not only continue to reap the consequences of unprotected sexual activity; we shall ensure that the next generation does too. In the vast majority of schools – as Schofield and countless others have pointed out – sex education is abysmally inadequate. If children are told anything at all about sex it is usually just the biological facts. And so we produce another generation of parents and teachers who not only don't know much about contraception or, more important, about the role of sex in the context of mature relationships; we produce another generation who are too ignorant or embarrassed to tell *their* children. Breaking this endless circle is perhaps one of the most important educational tasks facing all societies in the next decade.

These three attitudes to contraception – sex is for procreation, contraceptives mean more sin, and sheer embarrassment – go a good way towards explaining why good contraception is still so scarce. Given them, it is not surprising that so many people still have to be told that it is now possible to have a carefree sex life and not become pregnant.

Yet this is only half the picture. A large proportion of women (and men) who don't use effective contraception *do* know about it. What prevents their getting it is not so much their own lack of motivation as that of the people who have the power and resources to provide effective contraception: in short, society and its institutions. One has to meet an expert to get modern contraception, and we have failed to make his door sufficiently visible, reachable and affordable.

At one end of the scale, there is a hard core of 'problem' families who seem to be totally fatalistic about producing babies year after year. They are not so much feckless as socially inadequate. Harassed by the struggle to cope with life, they are caught in a vicious trap of poverty (and often religious dogma), incessant child-bearing, overcrowding and deeper poverty that puts them beneath the threshold of hope.

What they need is domiciliary birth-control.* In Britain, this began in 1959 with the pioneering work of Dr Dorothy Peberdy in Newcastle-on-Tyne, and was soon followed by others. These pioneer experiments showed that the vicious circle of ineptitude and procreation can be broken. 'Problem' families *want* effective contraception and can use it if they are shown how. But it demands time, effort and sensitivity.

These domiciliary programmes can have a huge impact, and pay off handsomely. In the London Borough of Ealing one doctor and a nurse provide a domiciliary service for 100 families. In the

---

* For a good review of British domiciliary birth-control, see the three chapters on it in J. E. Meade and A. S. Parkes, ed., *Biological Aspects of Social Problems* (Oliver & Boyd, London, 1965).

two years before they began in April 1967, the 100 mothers had had 145 babies and 6 miscarriages; in the first eighteen months of the service there were only 3 pregnancies. The service costs about £10 per family each year, while the immediate saving for the local authorities or state on each child not born because of it is almost £200 (maternity grant or allowance, hospital bed for delivery, welfare milk, health visitor, etc.). But though some local authorities have seen the point, many still refuse to start their own schemes until someone will 'prove' that they are worth while: in March 1971 there were fewer than forty domiciliary schemes in the whole of Britain dealing with a total of not more than 3,000 families.

Most women who need good contraception are easier to deal with. About half of them need minimum encouragement and will come quite long distances to large, central family-planning clinics where free (or heavily subsidized) advice and material are available. They need to be told where the clinics are – by the mass media, or by leaflets in welfare and health clinics – but they need little more than that. The other half need slightly more help. They are too busy caring for their families to go to central clinics, but can get to one in their neighbourhood. They also need more intensive information: posters, letters and so on telling them where to go and what they might expect when they get there. Given these triggers they will do the rest themselves.

Until very recently, hardly any communities had seriously tried to help these women (let alone the problem families, or the unmarried). Despite the rapid growth of clinics in all non-Catholic countries, family planning has not made itself *visible* enough.

One reason has been the almost universal blanket of silence over information: one rarely sees an advertisement or poster giving information about family-planning clinics. Birth-control advertising is still illegal in many countries. It was banned by British Rail until 1969. Even where it is allowed, the fear of causing offence ensures that the advertisements are hidden among all those others for surgical belts and acne cures.

But the main reason is that too many clinics and family doctors put women off by erecting frightening cultural barriers. Many doctors are still lukewarm, or even icy, about contraception. Even when they are willing to help, many do not go out of their way to encourage the shy customer who needs coaxing to talk about her (or his) problem. As Sir Theodore Fox, ex-editor of the *Lancet* and ex-director of the Family Planning Association, has put it:* 'Quite a lot of practitioners feel that what their patient does about contraception is none of their business. A few of them think it actually wicked, and many perhaps find it distasteful.'

Obviously, family-planning clinics are more cooperative, but many still make going for expert advice a gruelling experience for shy women. Some take full particulars of each applicant, including husband's income; others have been known to ask women in the crowded waiting-room to 'Take your knickers off and sit over there'. Even if they avoid these unnecessary excesses, most clinics fall a long way short of being easy 'shops' where a woman can just walk in and come out with what she needs. So a large number never 'buy' – or, having 'bought', never return.

These reasons go a long way towards explaining why in nearly all countries good – i.e. 'closed' – contraception is still predominantly a middle- and upper-class 'luxury'. Every survey of who uses what contraceptive method shows that most Social Class IV and V women use primitive 'open' methods while Social Classes I and II are now almost entirely on the pill or cap. Some optimists argue that given time, education, more resources for clinics, more liberal attitudes about providing contraception for all, every family and unmarried girl will have effective, modern contraception. But clearly there has to be an enormous amount of change and effort before that time comes and, given more liberal abortions as a backstop, no more babies are born unwanted.

Is there any chance of a technological solution that could bypass all this change and effort – a contraceptive method that is so reliable and yet medically safe that it could be released from medical control and bought across the shop counter as easily as bread and milk? Such a technological 'fix' would radically alter the whole birth-control – and population – picture.

*Theodore Fox, 'Family Planning and the Family Doctor', *Proceedings of the Royal Society of Medicine*, 59 (1966), pp. 1153–6.

There is a strong case for saying that we already have such a method in the pill and that oral contraceptives should therefore be freed from medical control and sold without any restrictions. As everyone knows, using the pill carries a slight but definite medical risk: chiefly that it will cause a possibly lethal blood clot (thromboembolism). Thorough studies in Britain* have shown that the annual death rate is 1 to 2 per 100,000 for pill-users in the 20–34 age group and 3 to 4 per 100,000 in the 35–44 age group, while the chance of being admitted to hospital (but not dying) because of a clot is about ten times higher for pill-users than for other comparable women (1 in 2,000 and 1 in 20,000 respectively). More recent studies have shown that the risks are roughly equated with the amount of oestrogen in the pill formulae: some high-oestrogen pills (between 100 and 150 millionths of a gramme of oestrogen per dose) are four times as risky as some low oestrogen formulae (50 millionths of a gramme) and have consequently been withdrawn from the market in Britain.

Yet despite these slight risks and the sensational publicity surrounding them, the pill is still the *safest* contraceptive known – because it is the most effective method of preventing pregnancy and therefore the risks of abortion or childbirth. Look at the rather surprising figures shown in Table 1.2, based on the best available data for the failure-rate and death risks of various contraceptives.

Dr Malcolm Potts, Medical Secretary of the U.K. branch of International Planned Parenthood, has calculated that the risks associated with taking the pill are equivalent to smoking a third of a cigarette each day for three weeks out of four (the period for which women take oral contraceptives). As he says, from the viewpoint of the health of a society it would be more justifiable to have oral contraceptives in slot machines and restrict the sale of cigarettes to a medical prescription.

However, there are counter-arguments of which the strongest is that keeping the pill on prescription ensures that every user must be seen by a doctor who might be able to detect signs that she is a pill risk. Without this check, limited though it is, there would be more deaths and illness from the pill.

TABLE 1.2  1,000,000 women aged 20–34 used the method for a year: this is what happens

| Method | Pregnancies | Deaths due to pregnancy and delivery | Deaths due to method | Total deaths |
|---|---|---|---|---|
| I.U.D. | 30,000 | 7 | unknown | ? |
| Pill | 5,000 | 1 | 13 | 14 |
| Diaphragm | 120,000 | 27 | 0 | 27 |
| Safe period | 240,000 | 55 | 0 | 55 |
| None | 1,000,000 | 228 | – | 228 |

After D. M. Potts, *I.P.P.F. Medical Bulletin* (October 1968). Note that these figures are for Britain, with a low maternal mortality rate. Where death in childbirth is more likely the pill is relatively much safer, and unprotected sex more lethal.

Despite the known risks, I predict that there will be growing pressure on governments to free oral contraceptives from prescription, and that in the near future this controversy will become one of the major birth-control issues.

The most important reason is that oral contraceptives are changing. Scores of research teams are looking for pill formulae that lower the risks and produce fewer side-effects but still do their primary job of preventing pregnancy. Unfortunately, there are signs that the two goals – effectiveness and medical safety – may be largely incompatible. We may have to buy safety with a greater chance of a 'mistake'.

The two main kinds of pill now available illustrate this. Most oral contraceptives use a *combined* regime of two sex hormones – an oestrogen and a progestogen – to give a triple defence against pregnancy. They suppress ovulation; they make the lining of the uterus unsuitable for an embryo to implant itself there; and they make the sticky mucus in the cervix hostile to spermatozoa. This makes them about 99·5 per cent effective.

In contrast, *sequential* pills deliver only an oestrogen in the early part of the cycle to suppress ovulation. In the latter part they deliver an oestrogen–progestogen balance. The aim is to cut down side-effects by using smaller hormone doses and by keeping closer to the natural shifts in the sex-hormone balance. But doing this leaves the suppression of ovulation as virtually the only line of defence. The instructions for use are more complicated, and unless followed exactly the pill may fail. So sequentials are less effective (about 98 to 99 per cent) than combined pills. On the other hand, they appear to produce remarkably few side-effects, and fewer risks to life or health.

The same goes for the much newer 'one-hormone' pills. These contain very small doses of a progestogen (for example, chlormadinone acetate), and are taken every day. The side-effects are very slight but the progestogen acts only on the cervical mucus. It is not yet certain what this means as regards effectiveness, but the signs are that it might lower it considerably below perfection (by pill standards) – perhaps to 97 per cent.*

The upshot of these developments is that oral contraceptives are becoming marginally less effective – though still much more effective than any other method – and more medically innocuous. Given also that abortions are becoming easier to obtain in the rare event of a pill failure, it is on the cards that women will swing heavily to using low-risk, less effective pills when there will be less medical justification than ever for keeping these pills 'closed'.

Nevertheless, for psychological reasons as much as for their medical disadvantages, the present generation of oral contraceptives are far from being a perfect solution. New *kinds* of birth-control techniques are needed, and are being developed. What are the prospects here?

The first thing to be said is that despite recurrent enthusiastic newspaper headlines, progress is bound to be extremely slow. In all advanced countries – the countries where virtually all contra-

---

* The first trials – J. Martinez-Manautou, *et. al.*, *British Medical Journal*, 2 (1967), pp. 730–2 – found a method failure-rate of only 0·2 per cent and a patient failure-rate of 2·1 per cent, which is excellent. But a follow-up trial – see J. Zanartu, *British Medical Journal*, 2 (1967), pp. 771–2 – found a 3 per cent *method* failure-rate while the British F.P.A. has warned doctors that this is the rate to expect.

ceptive development takes place – legislation to protect the customer of new drugs from adverse side-effects and health risks is now so strict that pharmaceutical companies have to launch their new products through a staggering thicket of animal and human trials. As a result the cost and time scale for drug development has exploded.

One recent estimate by Professor Carl Djerassi, president of the Syntex research laboratories and professor of chemistry at Stanford University, California, puts the cost of developing an entirely new chemical contraceptive from scratch to the final stage of clinical trials at around £7·5 million, with a time scale of no less than 17 years.* This is under American regulations, which are particularly strict on contraceptives, and which demand two-year trials in rats, dogs and monkeys, seven-year trials in dogs, and the start of 10-year trials in monkeys before any contraceptive can be given a full clinical trial in man. The bill for these trials is enormous: to screen 25 compounds that emerge from the laboratory in the hope that one will pass all the tests and end up a winner means using 12,000 rats, 2,400 rabbits, 800 dogs and 5,400 primates (at £400 per chimpanzee, £800 to £2,000 per gorilla) before any human volunteers can be called in. Not surprisingly, even the biggest pharmaceutical companies are finding that contraceptive development is a luxury they can easily give up.

As Djerassi suggests, probably the only solution to this *impasse* – which could well bring progress in contraception grinding to a halt – is to tolerate higher risks for chemical contraception and to introduce the concept of *conditional* approval for new types of pill, rather like the aircraft industry's Certificate of Provisional Airworthiness. This would slash the cost and time scale of development to levels which would rekindle the industry's enthusiasm, though at the price of making the pill-taking public the guinea pigs in the later stages of development trials. It is not a comfortable solution but it may be a necessary one.

Despite these limitations, it is not too early to start discussing the implications of the three main kinds of novel contraceptive

* 'Birth Control after 1984', *Science*, 169 (1970), pp. 941–51.

methods that are being developed. Some are already at a fairly advanced stage of development.

## Male pills

Many men insist on keeping control of birth-control, and many women have to rely on them. But the only means men have now are sterilization, withdrawal and the sheath. An effective male pill would help transform this often disastrous situation. It would also let women take periodic breaks from continuous use of the pill: husband and wife could share equally the responsibilities and medical risks of chemical contraception (though on the belt-*and*-braces principle, many couples might use male pills merely to double their defences against pregnancy).

Unfortunately male-contraceptive research is making disappointing progress. Many chemicals that prevent sperm production have been discovered, but so far they have all produced severe side-effects or lowered male potency and libido quite drastically. Until recently there were great hopes for immunological methods: injecting a man with antibodies that immunize him against his own sperm so that he never makes any or destroys any that are made. Research along these lines is still proceeding in various laboratories (as is immunization of females against spermatozoa) but is still a long way from producing safe, practical solutions.

## Reversible sterilizers

With one exception, every contraceptive method now available demands a positive decision to use it. Often people do not make this decision, because of ignorance, apathy, forgetfulness or even a desire to take a risk. This is especially hard for the unmarried: a young, single girl, however careful she intends to be, may meet sex unexpectedly and unprotected. Ideally, the decision to become pregnant should be the positive one. The rest of the time women (and men) should in effect be sterile.

The exception that approaches this ideal is the I.U.D., or intra-uterine device. Once in the uterus it can be left there until the owner wants to conceive, when it is simply removed. But I.U.D.s have severe disadvantages. They should not be used until a woman has had a baby; from 20 to 30 per cent of all women expel them from the uterus; failure-rates range from 2·2 per cent (large spiral) to 16 per cent (small bow); * there is a risk of infection; and in many women they cause bleeding and pain. One of the most extensive studies of the I.U.D. ever made – of nearly 24,000 women, mostly in the U.S.A. – found that for one reason or another nearly half the women concerned had stopped using the device by the end of four years.* People who argue for I.U.D.s as the 'threepenny solution to world famine' rarely urge their wives to use them.

Sterilization is no answer either, for it is irreversible in women and reversible only with difficulty in men (the best success-rate I can find is 57 per cent). This may change, but until it does, men and women will probably volunteer for sterilization (and perhaps should only be *allowed* to volunteer) only when they really mean 'for ever'. Even in these cases, sterilization can produce serious psychological effects. Apart from later regrets because more children are wanted, sterilization can produce castration anxieties, depression, depersonalization, frigidity, impotence or premature ejaculation – especially if the man is not sexually confident. If he is, then sterilization can work well for men (and women) and even enhance sexual libido by removing the fear of pregnancy.

The first truly reversible sterilizer will almost certainly be a single injection of a low-dose progestogen.† The hope is that a single shot will protect a woman for three months to a year with-

* Reported by Dr Christopher Tietze, of the Population Council (New York) at the 8th International Conference of the International Planned Parenthood Federation, Chile, 1967.

† These must not be confused with the present rash of trials of once-a-month injections. Until the end of 1969 at least, these were almost without exception injections of traditional, combined chemical contraceptives. Though some claim high effectiveness and few side-effects, none seems to offer a big enough gain over taking pills to warrant a monthly trip to the doctor for a jab – a trip that is surely very likely to be postponed every now and then, with obvious results.

out upsetting her normal menstrual cycle. To achieve this lasting effect the progestogen has to be released slowly. This is done by wrapping the drug in thousands of tiny wax-like pellets of different sizes that are absorbed at different rates. No one yet knows whether these micropellets can deliver a steady continuous dose sufficient to prevent pregnancy with a large number of women, or how long it will be before the last pellet melts and the woman has to go back for another injection.

Ultimately it may be possible to provide protection for years with implanted capsules. The first capsules may be unacceptably large, but in a few years there will probably be capsules the size of a match-stick or cigarette butt capable of delivering enough progestogens to protect a woman for years. Implanting a capsule under the skin of the buttocks would involve only a very minor operation. When she wants to conceive, the woman would either have the capsule taken out or have a progestogen-cancelling injection of oestrogen.

Assuming these reversible sterilizers are effective and risk-free, what consequences might they have? Their biggest impact might be to affect the *spacing* of births. As we shall see in Chapter 2, one of the major failings of sophisticated family planning is not so much that couples have more children than they plan, but that they have them sooner than they want. Changing this would have an enormous effect on population growth as well as on family happiness. Their other main impact will be on the unmarried. The advantages are obvious – though it might be a long and stormy time before most parents accepted that their daughters should have capsules implanted soon after puberty as a standard medical procedure.

## After-pills

All present contraceptives have a fundamental psychological weakness. As the American biologist, Garrett Hardin, has put it, they require 'the rational anticipation of an abrogation of reason – which affronts both the logician and the poet in us'. Consequently, many women meet sex unprotected, especially if they are young and unmarried. Others face accidental disasters, like a sheath coming off or a diaphragm shifting. For them the post-sex 'morning-after' or 'just-in-case' contraceptive would be an enormous blessing.

There are many possibilities. One main line of attack is to destroy the highly susceptible fertilized egg (zygote) during its six-day journey down the Fallopian tube and into the uterus, where it implants itself on the uterus wall, there to develop for nine months. Several specific anti-zygotic drugs have been found and tested in animals, but so far none has been considered safe enough for human use. Few doubt that a highly effective and safe one will be found.

Another very promising line of attack is to prevent implantation itself. As one researcher has put it, 'Instead of letting the sex hormones prepare a feather bed in the uterus for the egg, make them prepare a bed of nails.' It is the progestogen hormones that make the feather bed, and the idea is to pump in oestrogens to stop them. The most likely seems to be the synthetic hormone stilboestrol. This has been used on humans many times. In 1966–7, for example, Dr John McLean Morris of Yale University gave fairly high stilboestrol doses to more than a hundred women – mostly rape victims who would be legally eligible for abortions. There were occasional failures, and many of the women felt very sick. But McLean insists that the failures can be prevented by longer doses – perhaps a daily pill for four to six days after intercourse – and the sickness by anti-sickness drugs, or by taking the pills at meal times. Several other groups, including most big pharmaceutical companies, are hovering on the brink of human trials with other stilboestrol regimes and stilboestrol-like drugs.

What practical effect these after-pills will have depends very much on their medical safety and how often they have to be taken. If they must be taken fairly regularly, married women who expect intercourse regularly may prefer the continuous protection of traditional pills. If they relied on after-pills they would have to take *them* continuously, unless they could pin-point their fertile mid-cycle patch accurately. If one was enough, women might then prefer them. However, their main advantage will surely be

for the unmarried, or for the psychologically disturbed woman –
and there are many of them – who tries to conceive to hold a lover
or husband and then, a few days later, bitterly regrets her sudden
impulse.

But the greatest effect of all will surely be moral. After-pills
will stir up the centuries-old controversy of when 'life' begins –
when 'contraception' becomes 'abortion'. Most biologists insist
that the end of implantation, when the zygote is stuck firmly on
the wall of the uterus, is the earliest possible moment. In that
case, even a 'week-after' pill is 'contraceptive'. Religious views
are very mixed. In 1869 the Catholic Church ruled that the soul
appears at fertilization, so that any interference after that is
strictly condemned; yet, paradoxically, no Christian Church
baptizes or even demands proper burial of an aborted foetus less
than seven months old.

With that thought it seems time to cross the great ethical divide
into the tortured territory of the rest of birth-control – what
happens when contraception fails.

## Abortion

As long as contraception fails, vast numbers of women will go on
seeking and getting abortions rather than having to bear children
they do not want. They will do this whatever the law of the land,
the medical profession, or anyone else says about it. And contrary
to popular opinion, most of them will not be 'reckless' unmarried
girls but 'respectable' married women – for the simple reason that
most women of child-bearing age are married.

The great question facing practically every society is how far it
is morally and practically possible to open the doors to these
women so that they get their abortion legally and safely, rather
than illegally and dangerously.

In this section, I argue that the controversy must and will end
in a victory for permissiveness over restriction – even to the point
of abortion 'on demand'. It is now almost certain that biomedicine will make the controversy redundant anyway, by producing an acceptable abortion 'pill'. But even if not, the mere fact

that we now *know* more about abortion makes greater permissiveness almost inevitable. Then the important question becomes whether permissive abortion is possible in practice: can medicine and other social institutions meet the demand?

The starting-point for this prediction is the way the ground rules for the abortion debate are framed. They were laid down by Christian theologians around A.D. 200, but in modern dress go like this (from a Church of England report on abortion): *

> We have to assert, as normative, the general inviolability of the foetus . . . and then lay the burden of proof to the contrary firmly on those who, in particular cases, would wish to extinguish that right on the grounds that it was in conflict with another or others with a higher claim to recognition. Only so . . . can we maintain the intention of the moral tradition, which is to uphold the value and importance of human life.

This is an important statement. The right of the foetus to life should not be lightly overthrown. Yet however sincerely held, this statement is written from a defensive position. My point is that the higher claims that challenge the right of the foetus are bound to increase as our view of human beings and society enlarges and we see more of the evil consequences of *not* aborting.

The first challenge came when doctors learned that pregnant women with certain diseases risked death if their pregnancy continued.† This was a strong challenge in practice, because it meant a lot to doctors, and doctors carry out abortions. It has clearly been a strong moral challenge, too, for abortion to save a mother's life is now legal in virtually every non-Catholic Western country, and a growing number of Catholic doctors dodge their Church's ban on these grounds.

The second challenge built up as doctors learned of the risks of foetal damage from drugs and viruses. This was a weaker challenge. Risks are only risks and foetal damage is not always lethal.

---

* *Abortion: An Ethical Discussion* (Church Information Office, London, 1965).

† For a complete list of possible medical grounds for abortion, see 'Indications for Termination of Pregnancy' – a report by a committee of the British Medical Association (*British Medical Journal*, 1 [1968], pp. 171–5).

Morality and medicine could insist that any damage should be patched up when the baby arrived.

Despite this, an increasing number of countries now allow abortions for foetal risk. Britain, for example, made them officially legal in April 1968; and by the end of 1970 sixteen American states had done so too.

The third major challenge goes further and calls for the most liberal laws possible. It comes from many directions: from the growing demands of women for the right to control their fertility; from medical common sense – if women are going to get abortions anyway they might as well get them legally and safely; from cold medical economics – in 1964 some 35,000 women were admitted to hospitals in Britain after confessed illegal or self-induced abortions and took up 250,000 bed-days, or 80 times as many as the 2,000 legal hospital abortions took.

There are also strong pressures developing from the growing feeling that restrictive abortion laws make a mockery of the dignity of the law. Since these laws cannot suppress abortions they merely ensure that most abortions are illegal. This feeling has prompted even an American Catholic theologian-lawyer, Father Robert F. Drinan, to suggest that all abortion laws should be abolished.

Behind these various pressures there are two much more powerful forces at work. The first is that for many reasons advanced societies are becoming increasingly *child*-oriented. We seem to be heading for the day, predicted by the psychoanalyst Erik Erikson, when there will be a 'fervent public conviction that the most deadly of all sins is the mutilation of a child's spirit'. If so, when that day comes the very ground rules of abortion would have to be reversed: the right of the foetus to live would have to be proved higher than the inviolable right of every *child* to be loved and wanted.

The second force is the increasing evidence of the real and long-lasting harm that can come from *not* aborting. It was all there before, but it was scattered, anecdotal knowledge: what happened to friends or neighbours – how this girl lost her hopes of a career, how that woman had a breakdown when she was forced to abandon her baby, the child here who was rejected and 'went bad', the battered baby there. Such individual cases could be waved aside – and still are. But now that the human sciences are collecting the evidence and the mass media are thrusting it at us it becomes harder to ignore.

To present just one example of this new evidence, in 1966 two Swedish psychiatrists reported the first ever long-term follow-up of children who were born after their mothers had been refused abortions. There were 120 of them and they were all followed up to their twenty-first birthdays. Some of the results, compared to a matched group of normal children, were:

TABLE 1.3   What happens to children born after abortion is refused

|  | 'Refused' (percentage) | Normal (percentage) |
|---|---|---|
| Psychiatric consultation and hospital treatment | 28 | 15 |
| Registered juvenile delinquents | 18 | 8 |
| Public assistance after 16 | 14 | 2 |
| Education after legal minimum | 14 | 33 |
| Insecure childhood (fostered, or put in children's home) | 41 | 12 |

(From H. Forssmann and Inga Thuwe, *Acta psychiatrica scandanavica*, 42 [1966], pp. 71–88).

Clearly there is great and long-lasting harm here. If morality were based on reason and economics alone, restrictive abortion laws would soon be dropped. But then with the weakening of moral authority and the growth of all the pressures we have just looked at, are not these considerations just what the morality of abortion is beginning to be based on?

Now let us turn to practical issues. As many societies grope their way towards more relaxed laws, the important questions become whether it is possible to cope with them in practice, and whether everyone would be better for it.

Let me take the last consideration first – would we be better for it? – and pose the three key questions that are often raised to oppose more liberal abortions. Then we can come back to the practical question of whether we can cope with the consequences of more liberal laws.

## Is abortion medically harmful?

Overall, the safest abortions in the world are in those countries where they are easiest to get: Japan and the Soviet bloc. This is largely because most are carried out in hospitals, and at an early stage of pregnancy. When abortion is easy, few women have to spend those stressed and medically precious weeks searching for a sterile surgery and skilled hands. In Japan, for example, where there were 5,138,000 abortions in 1959–63, there were only 210 deaths – a rate of 4·1 deaths per 100,000 abortions. In Eastern Europe, where 'social' hospital abortions are not allowed after the first trimester, the rates (for the early 1960s) are 4·5 in Yugoslavia, 3·1 in Hungary and 1·2 in Czechoslovakia.

In other countries, death-rates can only be guessed at, because no one knows how many abortions there are. In Britain abortion deaths were running at about 50 a year until the 1967 Abortion Act, giving a rate of 50 deaths per 100,000 abortions if one accepts the usual estimate of 100,000 illegal abortions per year. This rate is more than ten times the Japanese figure.

As for morbidity, the main danger is permanent sterility. This can often occur after a botched back-street abortion, but is much less common in hospital abortions. A big Scandinavian survey* found that sterility follows from 1 to 6 per cent of legal abortions. This is roughly the same as for sterility following childbirth.

It is illegal abortions that are dangerous. Perhaps half of them are self-induced. The first resort may be gin and a hot bath, ergot, quinine or any of the scores of 'female remedies' sold over the chemist's counter. Because drugs rarely work, women may take large doses, and deaths from drug overdose, though rare, do occur. When the drug fails, they may go on to more desperate and dangerous remedies, such as the knitting needle or crochet hook. This is now rare in 'civilized' countries, though most women in

*J. Lindahl, *Somatic Complications Following Legal Abortion* (Svenska bokforlaget, Stockholm, 1959).

the world still abort themselves by squatting on the ground with a sharpened stick in their hand.

The more sophisticated back-street abortionist's method is to inject soapy water with a syringe. The great risk is of air or fluid getting into the veins of the uterus after the foetus has been removed. If air enters the circulation, rapid death is almost inevitable; if fluid does, it can cause gangrene or kidney damage.

The other main back-street killer is infection: the most dangerous organism, *Clostridium welchii*, can kill in twelve hours and is almost invariably fatal unless treated immediately by a skilled doctor. Other infections are common but rarely lethal, if treated, though a woman may suffer permanent damage to the kidneys, or to the uterus, Fallopian tubes and ovaries which may make her sterile. Antibiotics reduce the risk but they have to be given quickly. This means admitting the criminal abortion, which women are very reluctant to do, either because of their own shame, fear or guilt, or through an intense loyalty to the abortionist.

## Is abortion psychologically harmful?

The little evidence which exists suggests there is no connection between abortion and serious mental illness – psychosis or schizophrenia, say. An abortion will not cure it or cause it, and a refused abortion will not cause it either. Therefore the only important question is whether mentally-ill women should have an abortion or not. If not, it means forcing a mentally-ill woman to have a child she does not want – not a happy prospect.

With 'social' mental illness – neurotic guilt over an abortion, depression, stress reactions and so on – most psychiatrists agree that abortion is risky for two kinds of women: the narcissistic or masochistic woman, whose personal integrity might be destroyed by abortion, and the acutely depressed woman, whose inwardly-turned aggression will lead to overwhelming guilt if she has an abortion. Only prolonged and intense psychiatric support through and after the abortion – support which for obvious reasons can rarely be supplied – might offset the risk of suicide or breakdown if an abortion is performed: so the answer is usually 'no abortion'.

On the other hand, to refuse an abortion may also lead to suicide or breakdown. The frequently heard comment that no woman refused an abortion ever committed suicide is simply not true, as the case notes of many psychiatrists testify.*

With minor psychological effects, such as guilt and remorse, the evidence is very varied. In one Swiss study 50 per cent of women who had had abortions reported strong reactions of guilt and remorse. In a Swedish study about 15 per cent reported mild or serious self-reproach and 10 per cent extreme guilt. In a Norwegian survey only 2 per cent felt extreme guilt. Such wide variations suggest either the studies were bad or that the psychological response to abortion is socially conditioned. Though some Swiss hospitals are extremely liberal, the general climate in Switzerland is strongly anti-abortion. In Norway it is broadly permissive. In other words, a community which insists that abortion is psychologically damaging is not so much predicting grief as creating it.

## Is abortion socially harmful?

It is often said that widespread permissive abortion will lead to an erosion of the reverence for life, or even that abortion on demand is the first step to another Auschwitz.

Can this be so? If the charge is that *women* will become hardened by easy abortion this is a cruel distortion of their motives for seeking abortion. A few women may treat abortion casually, but most do not: they do not *like* killing the foetus inside them, either. If the charge is that society as a whole will become hardened there is no evidence that it has happened in societies where abortion is easy. On the contrary, the only society of recent times which tried to stamp out abortion so firmly that it made it a capital offence was precisely the one that did lead to Auschwitz – Nazi Germany in the 1930s. Permissive abortion laws stem from

* See for example some horrifying case histories in E. Tylden, 'Suicide Risk in Unwanted Pregnancy', *Medical World*, 104 (1966), pp. 25–8.

a more complex and less authoritarian view of human values and rights and are most unlikely to arise in or lead to rigid, inhumane societies.

## Can society cope?

If a country adopts liberal laws does it in fact lift the iceberg of abortion so that there are more safe legal operations and fewer dangerous illegal ones? If the answer is 'no', then abortion law reform on its own is largely a waste of time.

Broadly speaking, the answer *is* 'no' – unless there are two further major changes. First, the attitudes of gynaecologists – the people who do legal, safe abortions – must change to match the new spirit of the liberal laws. And second, if this does not happen, then the whole *structure* of the legal abortion system – where abortions can be performed and who performs them – has to change as well.

More liberal laws can only allow liberal gynaecologists to carry out a wider range of abortions safely and legally. They cannot force anyone against their judgement or conscience. Whatever the law, it is the gynaecologist who holds all the trump cards: the right to refuse, the right to make up his own rules for acceptance or rejection, the right to play God. And, for obvious reasons, gynaecologists are among the last people to accept the rationale of abortion on demand.* So what happens when liberal laws are introduced? There is an immediate surge in demand for the now legal hospital abortions, as more women learn that they can get

* However, there has been a radical change in worldwide medical attitudes since the late 1960s. Pre-1967 surveys of gynaecologists invariably revealed that while the majority favoured some liberalization of punitive abortion laws, only a tiny minority wanted to go as far as allowing abortion on broad 'social' grounds, let alone on demand. More recent surveys reveal much more liberal attitudes, culminating in the 1970 'Declaration of Oslo' by the World Medical Association. This was based on a British-sponsored resolution recommending that any abortion should be permitted when it is recommended by two doctors and performed by a competent doctor in approved premises. This declaration was approved by every national delegation but one (Argentina) and even by the Vatican Medical Association.

on the network through their family doctor without fear or stigma. Where there is a free health service, this surge is strengthened by 'old hands' switching from their private hundred-pounds-a-time contacts to the hospitals, and by 'abortion trippers' flying in. The hospitals become choked.

Then several things happen. The less liberal gynaecologists who have always refused the 'demand' cases now have even stronger excuses for doing so. The more liberal, who always managed to carry out hundreds of hospital abortions before the new laws, become overworked. And the flood of women who have to be turned away either go back home to have their babies, find someone in the back streets, or more likely get on to the private, hundred-pounds-a-time operators who have always been there but are now doing a roaring trade. Quite soon the system settles down to a new equilibrium, where there are more safe hospital abortions but otherwise with the pattern much as before. Most women will continue with the method they know – a visit to the helpful lady round the corner. It is much easier and quicker that way, no questions asked, and it is usually remarkably cheap.*

This surge followed by a drop, with only a slight reduction in illegal abortions, happened in Sweden and Denmark after they introduced liberal laws.† The initial surge, with long queues for free hospital abortions and a rapid rise in private 'abortion factories', was a key feature of British experience after the Abortion Act. When New York state introduced its almost unrestricted abortion on demand system in July 1970 hospitals were

* According to a 1966 National Opinion Poll survey of over 2,000 women who had had abortions, 63 per cent of those who had them illegally said they cost under £5 and a further 25 per cent said from £5 to £30.

† Sweden and Denmark introduced liberal abortion laws in the late 1930s, with further liberalization in Sweden in 1946 and Denmark in 1956. Before the Second World War there were fewer than 10 legal abortions per 1,000 live births in Sweden but after the 1946 laws the rate rose to 57 per 1,000 in 1951. Then it fell steadily and now varies between 27 and 31 per 1,000. In Denmark the peak came in 1961 with 70 per 1,000 and is now down to between 48 and 60. In neither country have back-street abortions dropped significantly, if at all. For comparison, the British figure of 84,000 legal abortions in 1970, the second full year after the Abortion Act, gives a rate of about 90 per 1,000 live births.

immediately swamped and it soon became quite clear that the system was largely excluding the poor.

What all this strongly implies is that abortion reform can never be completely fair and democratic unless it goes the whole way to abortion on demand and unless the structure of the abortion system is totally altered to allow it to meet the full demand.

With anything short of abortion on demand the situation will be very similar to that with contraception. Many women will not be able to face going through the machinery of approval for a legal abortion: the system will appear too forbidding and they will go down the street rather than confront society's well-dressed, inquisitive watchdogs.

To meet the full demand, one crucial step is to make it easier for the more liberal and sympathetic qualified abortionists to do more abortions with less fuss. In effect this means supplementing hospital abortions with abortion clinics staffed by qualified gynaecologists who charge minimal fees – enough to cover overheads plus a *small* charge for professional services.

Several such clinics have been set up in Britain, particularly in areas like Birmingham where the gynaecological top brass are strongly anti-abortion. But there are not nearly enough of them yet, as shown by the wide regional variations in the availability of N.H.S. abortions and the continual complaints about abortion patients overloading gynaecological departments of N.H.S. hospitals.

Obviously this cannot be the only step, as it is limited by the number of gynaecologists who are liberal enough to do abortions on demand and the natural repugnance of any gynaecologist at doing abortions day in, day out. Short of a massive change in attitudes from the profession, the consciences of all too many gynaecologists who do 'on demand' abortions will have to be bought, and bought dearly.

In that case, the answer is to lower the qualification standards. This is not such a desperate solution as it sounds, because there is a technological 'fix' that makes it possible – the vacuum method of abortion, in which the foetus is literally sucked out. The technique is so easy to learn that family doctors can use it perfectly well after a few hours' instruction. It does not demand any surgical experience and there is no need for the patient to stay overnight in bed. In fact in the U.S.S.R., Eastern Europe and China, where it is now the most popular abortion method, women are out of the clinic within an hour of their abortions. Vacuum 'lunch time' abortions have also become very common in New York since the July 1970 law allowed any abortion to be done by any qualified physician without any notification requirements or residency qualifications. The estimated 120,000 abortions in the state in the first year after the new law would hardly have been possible without a massive use of this quick, simple method.

The only drawback to its more widespread acceptance is that many gynaecologists think it slightly dangerous and a possible hazard to future fertility: in Britain, for instance, approved abortion centres have been warned by the Department of Health not to use the method. Others strongly disagree: one of the first doctors to use it in Britain, for example, has reported fifty suction abortions without any significant complications.* But whatever the risks, no one has suggested they are higher than with the average run of back-street abortions – the very abortions it could really help to reduce.

None of these technical solutions, however, gets to the heart of the problem with 'liberal' laws, which is the paraphernalia of psychiatric and medical assessment needed to ensure that the abortion is legal. As Scandinavian experience shows, this is the largest single obstacle to reducing back-street abortions.

Only when society allows abortion 'on demand' and provides the services to meet the demand will the situation change radically – as it has in the Soviet bloc. The aim of their laws† is quite

* Dorothy Kerslake, 'Vacuum Aspiration', in *Abortion in Britain Today* (Pitman Medical Publishing Co., London, 1966).
† In Bulgaria, Hungary, Rumania and the U.S.S.R. abortion is granted at the request of the woman. In Poland she has to establish her 'difficult social situation' in conversation with one doctor, who can then authorize the abortion. In Hungary there are Medical Boards to authorize abortion, but they are a pure formality since they must now by law agree to a request if the applicant insists on it. In Yugoslavia and Czechoslovakia authorization is by panels of doctors and representatives of the social services, but

explicitly to reduce the harm of illegal abortions and (to quote from the introduction to the U.S.S.R. laws) to 'give women the possibility of deciding for themselves the question of motherhood'.

These easy abortion laws have had three striking results. First, they have dramatically cut back-street abortions, though they have not eliminated them. In Czechoslovakia, for instance, there are now three hospital abortions to every one out of hospital, and in Hungary the ratio is five or six to one. Secondly, they have also heavily cut the birthrate, despite the fact that some of these countries have policies which strongly encourage childbirth, for example by large family allowances.

The third and more sinister result is that they have – or are said to have – produced an 'abortion mentality'. Instead of using good contraception, it is said, women use abortion as the preferred birth-control method.

There is probably some truth in this. In Hungary in 1964 there were 1,400 abortions for every 1,000 births; in many Soviet cities there are as many abortions as births; and in other Soviet bloc countries abortion rates vary between 250 to 600 per 1,000 live births.

These seem like staggeringly high figures, and are often quoted in the West as a powerful argument against allowing abortion on demand. Yet in fact they are probably not exceptionally high compared to a good many other industrial nations. For instance, it is estimated that in several Western European countries there are as many illegal abortions as live births. Even in Britain – with

---

their reasons for granting abortions are wide. (In Yugoslavia, if the birth of a child 'would result in a serious personal, familial or economic situation for the pregnant woman which cannot be averted in any other way'. In Czechoslovakia reasons include the woman's age, a 'large number' of children, the death or disability of her husband, likely disruption of the family, the woman being principal breadwinner, the woman being unmarried and facing a difficult situation if she bears her child.) East Germany is restrictive. Abortions are carried out in hospital where the woman stays usually two or three days. She has paid sick leave afterwards if she is a wage earner. Medical abortions are free, but those on request must be paid for, though the charges are heavily subsidized.

its long tradition of family planning and contraceptive services – the abortion rate is probably around 200 per 1,000 live births (200,000 abortions). In the U.S.A., according to the demographer Alan Guttmacher, there are probably 900,000 abortions a year, or 200 per 1,000 births.*

So, in a sense, every country has an 'abortion mentality', and the only real difference between West and East is that in the former the abortions are mostly illegal. Also, wherever one finds very high abortion rates one also finds that contraceptive usage has *always* been backward – not that it has once been good and then slipped back. Thus there is no real case for the 'abortion mentality' argument, and a very strong case for providing a good family-planning service with an abortion-on-demand service to back it.

However, it now looks as though science will reduce the need for providing either by inventing a safe, effective abortion pill.

Abortion pills are not the same as the 'morning-after' and 'week-after' pills now being developed, though these are often called abortifacients. Women will use after-pills as pregnancy *preventers*. They may act after contraception, but the *intention* of the woman taking them is not to let pregnancy start. A true abortion pill ends a *known* pregnancy. It is a true *tabs uteri evacuans*. A woman would take one after she has missed a period and confirmed that she is pregnant, or even when she has been pregnant for as much as five or six months.

The key to abortion by pill, possibly even on a do-it-yourself basis, is a remarkable range of chemicals called prostaglandins. Known since the mid-1930s as something of a curiosity, they were found in 1965 to be exceptionally powerful agents for making the uterus contract. Immediately they were taken up as a possible way of inducing labour at term.

Then, in the latter part of 1969, two doctors at the Makerere

*This estimate is not a wild guess but is based on figures such as that 12 per cent of urban women admit to at least one abortion. It agrees closely with preliminary estimates for New York state abortions in the first year following the July 1970 introduction of abortion on demand: taking state residents only, there were 198 abortions per 1,000 live births.

Hospital, Uganda – Sultan Karim and Marcus Filshie – used prostaglandins for the first time to terminate pregnancies. Using 15 patients between 9 and 22 weeks pregnant, they gave continuous intravenous infusions of prostaglandin F2-alpha until after a few hours fourteen of the women aborted (with the fifteenth woman the method failed). Side-effects were not serious but included fairly severe diarrhoea in seven women and vomiting in three.

When the results were published in January 1970 they caused a worldwide medical sensation. Medical researchers and pharmaceutical companies weren't slow to get the message that if a safe, effective prostaglandin abortion method could be perfected it would have revolutionary implications both in the developed and developing world, and hence enormous sales. The rush was on.

The first hurdle to overcome was to synthesize prostaglandins, which had previously been available in only tiny, very expensive quantities extracted from some corals and certain sheep glands. By the end of 1970 several firms had found suitable methods and were promising kilogram quantities during 1971 – one kilogram provides very roughly 3000 'abortion doses' by present techniques – and at a cost of only a few shillings per abortion.

As the drug became available during 1970, at least four large-scale human trials were begun in Britain, the U.S.A., Sweden and Uganda. By the end of 1970 they had produced well over 1,000 successful abortions. The main aims of these studies were to discover which types of prostaglandins are most effective, which have the least side-effects, and above all whether the cumbersome method of intravenous infusion can be discarded in favour of giving the dose by mouth or vaginal pessary.

At the time of writing, the answers to these crucial questions are not clear. Oral administration (by sucking a long-lasting pill) has worked successfully but the much higher doses required produce serious nausea and vomiting. Vaginal administration has proved more promising: in one experiment Dr Karim used one tenth the dose needed for the intravenous route for only two and a half hours and produced successful abortions with few side-effects. Equally significant, a small group of 20 women have been given a vaginal prostaglandin pessary once a month, with a total control of fertility – raising the possibility of an effective, do-it-yourself once-a-month contraceptive method.

It is not easy at present to predict just how large an impact prostaglandins will have on the abortion scene. At present their serious drawbacks – the side-effects, occasional failure to work, and fairly frequent failure to abort completely, so that the placenta and other material has to be removed surgically – mean that they have to be given under strict medical supervision.

With further improvement – and no one seems to doubt that this is bound to happen – abortions could be done on a semi do-it-yourself basis under loose medical supervision, perhaps at 'menstrual regulation centres' tied in to family planning clinics.

But if they are perfected to a point where oral or vaginal dosages are both effective and safe – and this is confidently predicted too – then prostaglandins will have an enormous impact, no less profound than that of the Pill. Abortion would become morally, psychologically, medically and socially silent – almost as unobtrusive as taking aspirins to stop a headache. No one would have to carry out abortions except the women who want them; and as Garrett Hardin has predicted, most women may actually prefer to use abortion pills than contraception.

For better or worse, there would be little point in trying to suppress abortion pills or pessaries for moral-religious reasons, or trying to impose the same medical control on them as doctors now apply to abortions. Even though there might be sound medical reasons for controlling the pills, control would be bound to produce a huge illicit black market in abortion pills, probably on an international scale. Abortion has always been an underground business because women have always wanted abortions badly enough to go to any lengths to get them. And when they know that cheap, safe, effective and self-administered abortions are available – whether legally or illegally – they are going to go out and get them.

# 2. POPULATION CONTROL

Birth-control is not population control. Birth-control depends on individual actions motivated by individual goals, not by community needs; and its central ethos is that parenthood is a private right, not a social privilege. The differences might not matter, except that in nearly every country men and women want more children than in the long term the community needs.

Population control therefore really does mean what it implies: control – from the subtlest persuasion to the strongest restraint – of private wishes and rights for the corporate good. This is a very far cry from the *de*control of unwanted fertility that is the true gift of effective birth control, making population control one of the most deeply resisted notions of our times.

Meanwhile, every day we hear of human masses living in misery. We read of teeming populations struck by drought and famine and disease, of millions of children stunted by malnutrition, of thousands of villages locked in poverty and underemployment, of their people streaming to cities already bursting and collapsing into squalor. Calcutta, they tell us, may grow in fifty years to contain sixty million people, and at present rates the population of the world's cities will in a century multiply forty times and the area they cover may increase a hundredfold to equal one-fifth of the entire land surface of the planet. Every year the figures tell the same seemingly relentless story: while the world's ability to produce food and other essentials slowly rises, the amount each person gets slowly falls. While the privileged third of humanity gets richer, the poor two-thirds gets steadily poorer. Nation after nation, struggling to break the deadlock of its poverty and subsistence economy, finds itself slowly sinking back under the sheer weight of its own numbers. The experts have started to predict not whether but when these countries will crash into famine and collapse – throwing not only themselves but the rest of the world into chaos.

Meanwhile in the privileged lands the struggle is not to survive but to survive decently without wrecking the environment. What we face are the problems of an affluence and mobility explosion piled on a population creep. It is not pressure of numbers so much as lack of planning and the lust for goods and services that make our cities decay and our suburbs sprawl, that jam roads, pollute air, water and soil, throw a thickening spatter of steel and concrete across the land, and kill year by year more of the quiet remote places. When the number of cars on the roads more than doubles in a decade, as it did in Britain in the 1960s, a 7 per cent population rise in the same period hardly matters to a road planner or anyone looking for an empty stretch of beach.

And yet, of course, it does matter. Even in slow-growth nations the laws of compound interest still work; and if they work for long enough they still create absurdity. Britain's 0·6 per cent a year population growth – which many experts call *stable* – still means a doubling in little more than a century, a mere one and a half life-spans away. Continued for 2,000 years it would multiply the population nearly a million-fold, giving 150 people per square yard.

Yet we don't have to look to such ludicrous distances: there can be very few people in Britain, or any other densely populated or highly suburbanized society, who can contemplate even one more doubling of numbers without the most appalling misgivings. The message is obvious. Some time, somehow, population growth in all countries, rich and poor, will have to be halted or even put into a steady decline.

Millions of words have been written about curbing population in the developing world, and though this is perhaps the most urgent and awesome of all human problems, I am not going to add to them here. Apart from having nothing new to say, population control in the rich, slow-growth countries – where the problem now is that women want more children than they should have rather than having many more than they want – is a far more interesting and contentious subject. It prickles with very real dilemmas.

The first, and absolutely basic dilemma, is how any society or its elected government can ever decide *when* it must act to curb growth when everyone assumes – as they do – that the curbs will be unpopular or at least upsetting.

This crucial difficulty has come up again and again in the recent wave of high level discussions on population that have taken place in Britain and the U.S.A. Sparked off by the growing concern over the environment, in Britain this new wave has been a truly remarkable reversal of years of ostrich-like evasions on population * – though so far it has been all talk, no action.

The main thread of almost every statement by every top government official in these discussions was a firm belief that though population growth would have to be curbed some time, there was no overwhelming case for saying that that time was now, or next year, or in the next decade. Indeed, given that

*The wave started in 1969, when Sir Solly Zuckerman, then the chief scientific adviser to the government, and his staff at the Cabinet Office, began confidential discussions with the relevant government departments at the highest levels into the need for and possible ingredients of an official population policy. In January 1970 the parliamentary Select Committee on Science and Technology started an exhaustive inquiry on the same lines. This was axed by the June 1970 election but was reconvened in December 1970. The final report of the committee, published in May 1971, came out in the strongest possible terms for government action to curb population growth. Its main conclusion was that 'the Government must act to prevent the consequences of population growth becoming intolerable for the every day conditions of life'. The report recommended the setting up of a special office directly responsible to the Prime Minister to coordinate and research population questions, and also to publicise the effects of population increases and their consequences. The report is a mine of information on the facts and attitudes among senior civil servants and ministers to Britain's population growth (*First Report from Select Committee of Science and Technology. Session 1970–71. Population of the United Kingdom*, H.M.S.O., 1971). Also in 1970 a group of Members of Parliament, worried by the rising tide of letters in their mail bags on over-population, started inter-parliamentary talks on the subject. Population was even due to be debated in the House of Commons in November 1970 but was shelved at the last minute by an over-full schedule of other matters: if the debate had taken place it would have been the first on population since before 1939.

population growth is comparatively slow and the problems of deliberately curbing it peculiarly intractable, it was hard to see how one could *ever* make out a watertight case for acting at *any* time.

This argument for inaction is almost inevitable from any elected government. For one thing, they are far too insecure to tell voters they shouldn't have more children (unless the voters clearly state that they also are worried by population growth) so governments have to convince themselves that population growth is the great unalterable and can be managed. But a deeper reason is that governments are elected to manage things and so surround themselves with hordes of planners whose main job is to come up with schemes showing that things can indeed be managed, population growth and all. Any doubts on this score can always be put down with vague talk about 'accepting change' and the need to adapt to new (that is, denser) environments, while the uncertainties of long-range population forecasts are a marvellous prop when you want to convince yourself that you don't have to look too far ahead in your planning.

To be fairer to politicians, the question of what aspects of a society or its environment will or will not be manageable in 30 or 50 years' time is so complex and so full of genuine uncertainties that even the most thorough analysis and expert advice cannot provide unequivocal answers. For the same reason it is impossible to give a rational answer to the question of what the optimum or 'best' population for a country would be. In the end the answers can only come from 'political judgement' – that is, hunch. And since taking positive steps to stop population growth is *not* an easy matter, the political answer is, not unnaturally, let's wait and see what happens.

If this *laissez faire* approach seems pathetically flabby, it is no worse than the strident noises made by the opposite camp – the fervent conservation lobby that demands instant policies to slash growth to zero and then put it into rapid reverse, all inside thirty years or so. Such shock tactics would create drastic changes in the age structure of the population. There would be a dramatic decline in the numbers of children, to be followed about twenty

years later by sharp declines in the working-age population and a
rise in the proportion of old people to the work force. Almost inevitably there would be a sharp counter-reaction, birthrates
would shoot up again, and the whole population structure could
go into a violent series of pendulum swings that would be socially
devastating. No less important – though the de-growth extremists
usually ignore it – any policy that was really going to bite into the
birthrate would also involve very severe and probably intolerable
restraints on the number of children any family could have. For
instance, if the United Kingdom population in A.D. 2000 is to be
roughly the same as the present figure of 56 million, average
family size would have to be slashed by about one child – from
2·4 to 1·5 children. Some conservationists have even demanded a
United Kingdom population of 40 millions by A.D. 2000 or soon
after – without realizing that to reach their goal there would have
to be a *negative* birthrate.

Yet the foolishness of some extreme demands to cut growth
does not mean that one ought to delay, as the politicians would
like. Any delay is only piling up trouble for the future. In Britain,
or any country where population is growing at less than 1 per cent
or so a year, all that is needed to reach zero growth is a gentle
touch on the brakes to reduce births. If this was done now it
would be 25–30 years before its impact was really felt, mainly by
lowering the number of future parents and hence the children
they will have. And it will do this without any drastic distortions
in the important age distribution of the population. The longer
we delay the more likely it is that when we do act it will be in
panic, with harsher curbs that will produce more violent, un-
stable changes – much like a skid when one stamps on the brakes
rather late.

What we should do when considering population control is to
examine as thoroughly as possible, taking a perspective of at
least 50 years ahead, which restricts our freedoms or quality of
life the most: our present non-decision to continue with un-
checked population growth, or the restraints needed to curb that
growth to any given amount. In this debate the widest public
participation is absolutely essential. Population and its control is

a political question, affecting every other aspect of social life, so it is totally unreasonable that answers on it should be taken solely by government committees, whether confidential or otherwise.

One would need a whole book to outline the first side of this equation – the costs of population growth – but the main effects can be summed up fairly shortly.

Looking to the year 2000, the population of the United Kingdom is now expected to rise to 66·5 million, or 19 per cent more than today's 56 million. Nearly all this increase will be in England (with Wales, up from 48·9 to 58·6 million) though England is already one of the most crowded countries on earth. The heaviest pressures will be in the north-west and south-east regions, where nearly half the population is now packed in at densities 2·4 and 1·8 times the national average. Indeed, the pressure on the north west is so great that on present trends, well before the end of the century it will have run out of room and will be able to develop its industry and housing only by climbing up the Pennine and Welsh hills.

Consider some of the stresses implied by this growth allied to rising consumption and standards, remembering that most social needs are grossly under-catered for already. The car population and demand for countryside recreation are both expected to triple by A.D. 2000: peak holiday areas such as Cornwall and the Lake District are already saturated on fine summer weekends, and strict rationing – even to the extent of policed cordons to keep further crowds out of vast holiday areas – have been seriously forecast.

The under-fifteen population will jump in the next 30 years from 11·5 to 14·3 million, an average increase of 2,000 each week. The equivalent of two 650-child schools must be built each week for the next 30 years to take these children before any attempt can be made to deal with antiquated or overcrowded conditions, over-large classes, or the dire lack of nursery schools for the pre-school child. The extra 10 million population will need houses equivalent to building 30 Nottinghams, even if there are no further improvements to today's dismal housing standards. Urbanization and industrialization are expected to bury an area equivalent to Cornwall and Devon under concrete and asphalt. Farm yields must rise by 80 per cent if we are to continue official policies of growing half our own food: this may not be possible without disastrous 'over-mining' of the soil and other fundamental ecological changes (despite prodigous fertilizer use, the rising yields of the last two decades are already falling off).

Nor is this just *our* problem. Each Briton consumes in a lifetime something like 20 times as much of the earth's finite resources of minerals and fossil energy as the average Indian, and pollutes on a similarly gargantuan scale. By this measure our population of 56 million is already more of a global disaster than India's 500 millions. To say the least, it is unwise to assume that this gross imbalance can continue indefinitely.

Enough. The economist, Maynard Keynes, once asked: 'Is not a country over-populated when its standards are lower than they would be if its numbers were less?' There is very little doubt how most people, even high government officials, would answer this question. As far back as 1965, before the environment became a central issue, a large survey * found that 73 per cent of people interviewed in Britain believed a larger population would be a 'bad thing' for the country. It is inconceivable that the proportion will not have risen since then, though as far as I know there have been no similar surveys. Yet, very significantly for the main theme of this chapter, a large majority rejected any idea that population should be curbed by government impositions.

Now let us look at the other side of the equation. If a majority want an end to population growth, what 'government impositions' would actually be involved in bringing it about? To what extent does population control mean curbing our 'right' to have as many children as we want – a perfect excuse for governments to avoid action – and to what extent does it mean *de-control* by setting up excellent family planning and abortion services to help everyone not have babies they do not want? And if it does mean the first, to what extent would anyone really mind?

* *Worldwide Opinions about Some Issues of Population Control*, United States Information Agency, December 1965.

Before reaching any answers, we must look at the key demographic facts in Britain. The most striking factor in the population picture since the war has been the way birthrate climbed steadily to a peak in 1964 of 18·5 births per thousand population (for England and Wales), and has since dropped steadily to 16·3 in 1969 and 16·0 in 1970. At the same time, official forecasts of the population at the end of the century have swung up, but are now falling as birthrate drops: see Figure 2.1.

FIGURE 2.1

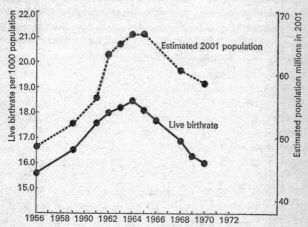

Forecast population in England and Wales 2001 and live birthrate in recent years. Trends for the United Kingdom are very similar.

Since the death rate has remained remarkably steady for the last few years at between 11 and 12 deaths per 1,000 population – and in crude terms it is the excess of births over deaths that produces population growth – many people have started saying that the population problem is well on its way to curing itself (see Fig 2.2).

This is hardly so. To achieve zero population growth in the long run, Britain's birthrate will have to fall to around the 14 mark, when the population would be exactly replacing itself each

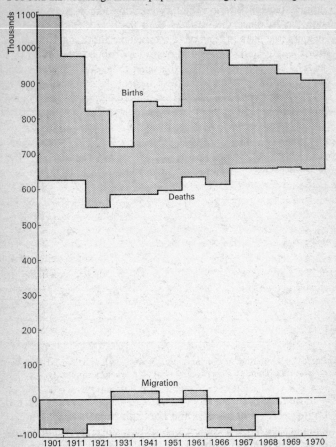

FIGURE 2.2 Average annual population change, United Kingdom

year. The latest birthrate lows of 16·3 and 16·0 in 1969 and 1970 are still a good way above this figure. Indeed, as the Registrar General's projections warn, the gap still gives a picture of 'substantial growth'. In fact all the downward trends in birthrate since the 1964 peak merely amount to a delay of 12 years – from 1992 to 2005 – for the date when England and Wales will reach a population of 60 million.

However, birthrate is a slightly misleading figure. Now that most families are fairly uniformly small, what really matters in the long run are changes in average family size: so much so that this is now the critical figure on which assessments of future population are based. This is hard on demographers, for, as Table 2.1 shows, it takes a long time for the average family to complete itself. Though most children are now bunched up into the first decade of marriage – allowing some quite informed guesswork at the 10-year mark about how many children there will eventually be from all the marriages in any given year – one cannot be absolutely sure until about 30 years have passed, when one can assume that all the wives are through the age of childbearing. To make things worse, future population size is acutely sensitive to small changes in completed family size.

TABLE 2.1  *Average number of children born in marriage (Great Britain)*

| Year of marriage | Number of children born after | | | Completed family size |
| --- | --- | --- | --- | --- |
| | 5 years | 10 years | 20 years | |
| 1931 | 1·14 | 1·66 | 2·11 | 2·14 |
| 1936 | 1·06 | 1·64 | 2·04 | 2·05 |
| 1951 | 1·20 | 1·87 | (2·23) | (2·23) |
| 1961 | 1·48 | (2·12) | .. | .. |

(Figures in brackets contain a small element of guesswork)

The important thing is what average family size is going to do next. Current estimates, based on the most thorough analyses of recent trends, suggest that having risen appreciably since the 1930s and dropped slightly from 1964 to 1970, family size will stick at around the 2·4 mark. That is, couples marrying in the

1970s are expected to have an average of 2·4 children, though some of the children may not be born until the very end of the century.

It's on this assumption that present forecasts of a U.K. population of 66·5 million in the year 2000 are based. What would happen, though, if families did not reach the 2·4 mark? According to the statisticians who make Britain's official population forecasts, if family size dropped to 2·25, the population at the end of the century would drop by 2·5 to 3 million (that is, to 63·5 or 64 million); and if there was a real slump to an average family of 2·0 children, there would be a further drop by 2·5 to 3 million (that is, to 60·5 or 61·5 million, compared to the present population of 56 million). This last change, down to the 2-child average family, is considered to be the most extreme that is possible – short of some very powerful new factors entering the scene to change fashions drastically, like a major drawn-out slump, a really deep and universal concern about over-population (or an aggressive population policy). As Table 2.1 shows, the only period when families were so small was during the depression years of the 1930s.

However, though the 2·0-child average would mean that population continued to grow up to the year 2000, in the long run it would produce a steady population decline. Most of the increase this century would be due to higher birthrates in the past producing a 'parent bulge' over the next 30–40 years. When this had worked itself out, birthrates and population growth would fall steadily. In fact for Britain the critical 'breakeven' family size is 2·1 children. This is the size which over the long run would produce exact replacement of the population; or in other words, zero population growth. (The extra 0·1 child is needed to offset deaths before the age of 50.)

At present, though, we seem to be set on having 2·4 children per family. Yet this is only an average. Very significantly for any hopes of reducing population growth, this average conceals a wide spread of family sizes when one breaks it down for different ages at marriage. The official forecasters reckon that for marriages in the 1970s, where the bride is under 20, there will be 2·88 children in the average family; for 20- to 24-year-old brides, the assumption is 2·3 children; for 25- to 29-year-old, only 2·0.

Thus young brides are a major target for population-control measures. Interestingly, they are expected to have more children not because they have a longer time 'at risk' in which they could have an unwanted child or two, but because the extra years of marriage give the subfertile a greater chance to conceive. Young brides, in other words, are less likely to end up childless. Also – a very important factor – young brides tend to *want* more children; conversely, older brides tend to want fewer, often because they have had time to find other interests outside the home. As most people know, there has been a marked swing in the last 40 years towards younger marriages: from 1931 to 1968 the average age of first marriage (for women) dropped from 25·7 to 22·7. A large part of this swing is accounted for by teenage brides: in 1931 about 10 per cent of women married when less than 20, today about 30 per cent do.

With those outline demographic facts behind us, we can start to look at the various targets for any policy designed to reduce population growth.

An obvious first target – so obvious that for many it is the only *necessary* component of a population policy – are the tens of thousands of unwanted children who are born each year. Though some rather wild estimates put these as high as 250,000, or over a quarter of all births in the United Kingdom, it is wishful thinking to suppose that even the total elimination of all unwanted births by some magically effective birth-control campaign would in fact result in the disappearance of the population problem.

Consider illegitimate births for a start. They now account for 1 in 12 of all births, or 67,000 babies in England and Wales in 1969 (a slight drop on previous years). They look like a golden opportunity for a painless form of population control through better family-planning services. But despite popular myth, this is simply not so. Even the complete elimination of all illegitimate births would, the eminent demographer Professor David Glass estimates, reduce long-term population growth by only 2 per cent or so.

The reason is that nowadays practically all women – the figure for Britain is actually 92 per cent – get married within the child-bearing years. By so doing they not only legitimize any illegitimate children but bring them within the fold of the 2·4 children of the average family size. In other words, the average mother who has an illegitimate child will end up having had one child outside marriage but only 1·4 during her married life. Besides, for various psychological reasons, very many unmarried girls want or need to have an illegitimate baby, so that it might not be all that easy to eliminate illegitimate births.

However, there is no denying that more effective birth control would prevent countless 'shotgun' marriages with their exceptionally high divorce rates – a vast benefit – and would thus help delay marriage and childbearing. As we have seen, older brides have fewer children on quite a dramatic scale.

What about unwanted children inside marriage? This is not easy to answer, mainly because it is so difficult to make the surveys. If one asks the parents of, say, five children how many they wanted they will almost invariably say five, unless they are really deeply over-burdened with too many children. Hardly any can bring themselves to deny an existing child. The only hope of getting near a true answer is to find out the couple's plans at the start of marriage and then check them against actual production. But, as we have seen, this takes a lot of time to give results, and anyway parents revise their targets as babies arrive and their circumstances change. Another technique is to ask only about wanted or unwanted *pregnancies*, as women are more objective about them than they are about living children.

There is also the problem of defining 'want'. In a perfect world, with no problems about money, housing, education and so on, most couples would like quite a lot of children. Usually surveys find that this 'ideal' family size is at least one child up on average over actual completed family size, suggesting that deep inside them most women have another child bursting to get out. Putting it another way, when women are asked how many children they *intend* having they nearly always give a lower number than for the 'ideal', implying that there is a self-denying factor in family building closely tied to the flux of changing circumstances. At the very least, all this makes 'want' and 'unwanted' very hazy concepts.

Nevertheless, surveys consistently show that a large proportion of women – ranging up to 50 to 60 per cent – say they have had 'accidental' or 'unplanned' pregnancies. Though many of these mistakes are a matter of spacing out one's children – the pregnancy is unwelcome because it is too soon – there is a definite and fairly large hard core of frankly unwanted births.

Probably the best estimate of numbers in Britain has been made by Dr Ann Cartwright of the Institute of Community Studies.* Dr Cartwright surveyed nearly 1,500 women around England and Wales after they had had a baby and asked them searching questions about birth control practices and their attitudes to their pregnancies. The sample of women, moreover, was closely representative in terms of social class and other factors of the actual population. She found that for their most recent baby, 15 per cent of mothers were 'sorry it happened at all'. For all pregnancies, the 'sorry' figure was 14 per cent. As one might expect, the more children a woman had the more likely she was to say her latest pregnancy was unwanted. Where the birth was the third baby in the family, 19 per cent of mothers were 'sorry'; for the fourth, 25 per cent; the fifth, 39 per cent, and for later births, roughly 50 per cent. Table 2.2 gives fuller details, including the proportions of happy successes.

These figures are a terrible indictment of the effectiveness of family-planning services (and/or women's reluctance to use them) but they do not suggest that if these unwanted births were avoided the population problem would go away. The main reason is that very few people have very large families: only 7 per cent of families have five or more children and less than 2 per cent have seven or more. Huge 'feckless' families are not a significant factor in population growth in countries such as Britain.

Indeed, my own analysis of Dr Cartwright's findings combined with the known proportions of families which are of any given

* *Parents and Family Planning Services*, Ann Cartwright, (Routledge & Kegan, Paul, 1970).

size shows that if none of these unwanted pregnancies had happened, average family size would not drop below 2·27 from its present 2·4. The estimate can only be rough because one cannot know how many of the women who did not want a particular pregnancy might have wanted a child later. Yet this does suggest that eliminating unwanted births cannot be the whole answer, even supposing it was practical in the foreseeable future.

TABLE 2.2  Attitudes to pregnancy

| | *The pregnancy resulted in the mother's:* | | | | | | | | |
| | 1st | 2nd | 3rd | 4th | 5th | 6th | 7th | 8th | 9th or later child |
|---|---|---|---|---|---|---|---|---|---|
| | % | % | % | % | % | % | % | % | % |
| Preferred earlier | 6 | 7 | 4 | 4 | 2 | 2 | 2 | — | — |
| Pleased pregnant then | 64 | 69 | 60 | 53 | 47 | 36 | 46 | 44 | 44 |
| Preferred later | 22 | 19 | 17 | 18 | 12 | 10 | 5 | 6 | — |
| Sorry it happened at all | 8 | 5 | 19 | 25 | 39 | 52 | 47 | 50 | 56 |

As well as the absolute size of families, one must also consider the *timing* of births: not only when people get married but how they space their children. A staggering proportion of marriage- and birth-timing is not by free choice, while fashions can produce quite large swings either way.

In many countries, including the U.K., Denmark, Sweden, Germany and the U.S.A., roughly 40 per cent of brides under twenty-one are pregnant. Much of this probably just comes from not waiting for the honeymoon, but a great many of these marriages must still be based on some-love-plus-an-accident or on an-accident-plus-a-parent's-shot-gun, and with often bitter results. If all these girls had perfect fertility control many would probably delay marriage for years. Once married, we are still highly accident-prone. Most couples have their first baby as soon as possible. They do not use contraception to delay it, so that whenever it comes it is, in their terms, planned. Some couples go on like this until they reach their target, and their children are therefore also all 'planned'. Most couples, though, try to space things out, but try rather half-heartedly. As a result, according to several surveys, more than a half of all pregnancies that occur *after contraception has been started* are unplanned.

Since no one has measured how much too early these accidents are, it is impossible to estimate what effect perfect timing would have on the birthrate. What we can do though is to see what an astonishing difference quite minor timing changes make.

Imagine two countries. The first is inhabited by the Fast Breeders. Every girl gets married on her twentieth birthday, tries to have a baby straight away and has it when she is just twenty-one. She has two more children, spaced out at eighteen-month intervals. In the second country, the Slow Breeders get married on their twenty-third birthday – just three years after the Fasts. They use contraception for eighteen months and so have their first baby two and a half years after marriage. Their two later children are spaced out two and a half years apart. Neither pattern is anything like extreme for today's families, yet other things being equal, the birthrate of the Slow Breeders will be only 80 per cent of that of the Fast Breeders. And after ninety years, while a hundred Fast Breeders will have produced 504 *great*-grand-children, a hundred Slow Breeders will have produced only 336 grandchildren.

Shifts as big as the gap between the Fast Breeders and the Slow Breeders have already occurred in many developed countries in this century. For example, in the U.S.A. the average age for first births fell from around 24 up to the 1930s to $21-21\frac{1}{2}$ in the late 1950s – a drop of $2\frac{1}{2}$ to 3 years. For later children the drop has been even greater – roughly 3 to 4 years since the First World War. It is now known that this trend to earlier babies (and marriage) accounted for at least half the massive baby boom after the Second World War, the rest coming from a trend to larger families. More important, demographers are now guessing that the trend is being reversed. The recent big drop in the U.S. birthrate is probably not due to the pill (there was an equally steep drop in the 1920s) but to a swing to later marriage, more spaced-out children and perhaps also a desire for slightly smaller families.

No one really knows what sets these timing patterns. Incomes and the cost of living are obviously important – major slumps slow things down. Housing difficulties, more higher education and careers for women delay marriage and first children, but much of this part of the pattern is set by fashion: when one's friends are marrying young and getting on with starting families (and parents are breathing down one's neck) the pressures to conform are heavy. Child spacing is just as complex, with the wish to get through the baby stage quickly and have the children in school counteracting the mother's inability to cope with a clutch of babies and toddlers and the growing recognition of the emotional needs of the child. The spread of the Spock syndrome, with its message of how much individual attention infants and toddlers ought to have, could have a big impact on family spacing.

But if one could alter these factors (or they altered themselves), so that there was a fairly big shift from fast to slow, birthrates would drop substantially, but not by anything approaching the 20 per cent of the calculation above, unless the swing was really massive. The reason is that in any actual country there are always a good many slow breeders or even very slow breeders. For example, in England if two-thirds of all mothers up to the age of thirty delayed their babies by three years – in other words, switched to the reproductive patterns of today's university-educated couples – the birthrate would drop by a significant but not decisive 6 per cent due solely to the slower rate of breeding. But the total effect would be considerably greater owing to the tendency of older brides to have markedly fewer children.

Most demographers believe that what all these figures add up to is an almost certain indication that better birth control will not by itself bring population growth in Britain down to the zero mark – let alone produce a steady decline in numbers that many people would like to see.

There are too many unknowns to be absolutely sure. For instance, there is fairly strong evidence that better birth control itself leads to a desire for smaller families. When women learn how to plan and space their children they also take up roles outside the family, see alternatives to motherhood, and so come to prefer fewer children when they eventually start having them. But there is plenty to offset this population-reducing trend. Nearly 10 per cent of families are childless, not because the couple want to be but owing to subfertility or infertility. The growing power of medicine to cure these complaints could make a large difference. Fashions in family size could always swing upwards again, while it should never be forgotten how much population would continue to grow even if family size suddenly dropped now to the level for exact replacement – a legacy of higher birthrates in the last 20 years. In the United States this 'backlog growth' would continue to A.D. 2035 and increase population by 40 to 50 per cent to around the 280 million mark; in Britain the effect would be slightly less. Lastly, perfect fertility control through ultimately effective contraceptive and abortion services is not likely to be won overnight.

So if we want to stop population growth we will almost certainly have to go beyond family planning to deliberate policies aimed at reducing the number of children people *want*.

As Table 2.3 shows, policies would not have to produce vast changes in family size. The differences between the top rows of figures and the bottom rows – that is, between the actual pattern of family size today and a possible pattern that would produce

TABLE 2.3

| *Family size pattern in England and Wales today* | | | | | | | |
|---|---|---|---|---|---|---|---|
| Children in family | 0 | 1 | 2 | 3 | 4 | 5 or more | TOTAL |
| % families with this number | 9 | 18 | 34 | 20 | 12 | 7 | |
| So every 100 families has this many children | 0 | 18 | 68 | 60 | 48 | 46 | 240 children Average family size = 2·4 |
| *A pattern for zero population growth* % families with this number | 12 | 21 | 37 | 17 | 9 | 4 | |
| So every 100 families has this many children | 0 | 21 | 74 | 51 | 36 | 26 | 208 children Average family size = 2·1 approx., i.e. exact replacement |

zero growth – are never larger than 3 per cent, though they do involve large proportional adjustments. For instance, the number of families with five or more children would have to be nearly halved. Of course, the future pattern can be made up in many ways and it makes an interesting game trying to pick plausible patterns: the only rules are that the percentages of families must total 100 and the numbers of children must add up to near 210 (the exact replacement figure for 100 couples, or 200 parents).

Some experts, however, would say that figures like these of family size are a total irrelevance. They insist that however close or far a population is from replacement level, policies to reduce *wanted* children must be the main priority, while all talk about family planning as the big white hope for population control is a dangerous diversion. Because family planning aims to help people it is politically and socially acceptable; because it helps to reduce the birthrate by preventing unwanted children it appears to help the goal of damping population growth; but – the radical experts argue – because family planning studiously avoids any suggestion that wanted children might have to be 'prevented' too, it is a convenient excuse for avoiding the real issues, while seeming to cope with them.

The strongest attack on these lines was made at the end of 1967 by Kingsley Davis, Professor of Sociology and Director of International Population and Urban Research at the University of California, Berkeley.*

Davis condemns the evasive attitudes of developed as well as under-developed countries. Basically, what he says is that population policies should not be written off as failures, because they have never been tried. Though contraceptive and family-planning programmes are important, to call them population control programmes is to behave like an ostrich:

There is no reason to expect that the millions of decisions about family size made by couples in their own interest will automatically control population for the benefit of society. On the contrary, there are good

*'Population Policy: Will Current Programs Succeed?', *Science*, 158 (1967), pp. 730–39.

reasons to think they will not do so. At most, family planning can reduce reproduction to the extent that unwanted births exceed wanted births. In industrial countries the balance is often negative – that is, people have fewer children as a rule than they would like to have. In underdeveloped countries the reverse is normally true, but the elimination of unwanted births would still leave an extremely high rate of multiplications.

Davis argues that with its obsessive emphasis on the family and its health and needs, the family-planning ethos hinders the hope of population control in another, crucial way. It tends to protect or even boost the traditional image and rights of the family as a unit – a separate unit in charge of its own destiny. The only hope of curbing population growth is to shatter this image. To get an effective population policy, society will have to invent a complex, radical and perhaps painful set of rewards and punishments to change the motivation for having children. Keeping population control – or rather the distribution of contraceptives – in the hands of doctors and public health workers is hardly going to help us here. One result of doing so is that after two decades of realizing that we ought to know the answers, we still do not know why families choose to have as many children as they do, and under what circumstances they might wish to have more or fewer. No governments and very few private foundations are doing the kind of surveys and thinking and social experiments about population policies that are needed, because we have blithely believed that family planning would cure rather than just ameliorate the disease.

Another result – perhaps the most important – is that the 'hands off' attitude to tampering with family 'rights' has hardened, despite the vast growth in the acceptance of social planning in almost every other aspect of life. And so the idea of controls looks all the more horrific:

Social reformers who would not hesitate to force all owners of rental property to rent to anyone who can pay, or to force all workers in an industry to join a union, balk at any suggestion that couples be permitted to have only a certain number of offspring. Invariably they interpret societal control of reproduction as meaning direct police supervision of individual behaviour. Put the word *compulsory* in front of any term describing a means of limiting births – *compulsory sterilization, compulsory abortion, compulsory contraception* – and you guarantee violent opposition. Fortunately, such direct controls need not be invoked, but conservatives and radicals alike overlook this in their blind opposition to the idea of collective determination of a society's birthrate.

The great problem of 'collective determination' – the central dilemma facing any population policy – is of course how to invent controls that are effective and attractive (and which will not bankrupt the state). There are plenty of effective, unattractive controls. Most of these are at work all the time now, but for a government to foster them as population-control tools (or for any other reason) would be political suicide. As Davis says, as deliberate population manipulators they read like a catalogue of horrors:

Squeeze consumers through taxation and inflation; make housing very scarce by limiting construction; force wives and mothers to work outside the home to offset the inadequacy of male wages, yet provide few child-care facilities; encourage migration to the city (city birthrates are lower) by paying low wages in the country and providing few rural jobs . . . [And so on.]

But there are alternatives. Without the last resort of compulsion and legislation there are many beneficial or neutral ways of manipulating today's educational, economic and tax policies, first to delay marriage and second to limit child-bearing in marriage. Here is a list of the most important.

*Benefit controls*

Bonuses for voluntary sterilization; all abortion costs paid by the state; subsidized contraception for those needing it and easier access to present 'closed' methods like the pill (see Chapter 1). With better school education in sex, birth-control, etc., these are the main new factors to add to current family-planning programmes.

More higher education for all.

Stop taxing single people more heavily than married.

Large marriage bounty from the state if a first marriage is after a certain age (and the bride is not pregnant!). There could be a sliding scale here up to a maximum at, say, twenty-five. The costs of a sufficiently attractive bonus would be high – there are some 350,000 first marriages a year in the U.K. – but this is only 'play money' that would rapidly circulate and be recovered in other taxes.

Encourage wives to work by abandoning taxes that discriminate against working-wife couples. This could have disastrous effects on existing children unless there was a gigantic day-nursery programme also, which would encourage child-bearing. So, more radically:

Equal higher education, job opportunities and pay for women, plus school education to re-define the sex roles etc., in an attempt to shift women from their role as housewives and mothers to new interests outside the house and family. This is a major factor in the low Eastern-European birthrates.

Anti-family allowances: i.e. bonuses or tax relief for childless couples.

The main aim of this approach, Davis suggests, should be to lessen the emphasis on the family and provide attractive alternatives. However, though this may reduce population growth we have to be on guard against other, less desirable social effects. Some form of family structure is the core of all societies. Where deliberate attempts have been made to loosen the structure no very satisfactory alternatives have been found. The Israeli *kibbutzim*, for example, are still trying to replace most of the traditional family and child-rearing roles; but the children they produce very often reject this way of life in favour of old family patterns. A similar fate usually ends the many other types of experiment in communal living in other countries. Perhaps this is not surprising. Though many psychologists decry the nuclear family as a stressful setting for child-rearing, there is a good deal of evidence to show that no form of group care yet invented does as well in fitting children for life in society as we know it.

A simple answer would seem to be to concentrate only on the benefit controls that delay marriage or child-bearing, educating everyone into assuming that women will have a longer period of education and earning their own living before taking up the 'normal' wife and mother roles. Unfortunately, as we shall see in later chapters, this would be medically dangerous. Somehow we shall have to balance out the population advantage of older mothers with the genetic and other medical advantages of younger ones.

At the same time, since no state is infinitely rich, and some of these innovations would cost a great deal, they will probably have to be offset by more restrictive controls that bring some money back to the state. These are the ones that are usually talked about, and they carry great dangers. They either erode important freedoms or threaten existing large, poor families. Here is a list of the most important that have been put forward.

## Restrictive controls

All illegitimate pregnancies to be aborted. More humanely, intense pressure for voluntary insertion of I.U.D.s or other reversible sterilizers when available after an illegitimate birth or abortion.

Big boost for conscripted (or voluntary) overseas aid programmes for school or university leavers. Overseas military service is a powerful marriage delayer.

High charges for marriage licences. (This discriminates heavily against the poor and would be all too easy to avoid by simply not getting married.)

Stop special grants to married students or students with families.

Students would repay the state for their higher education when completed, say by heavy salary deductions for three or four years. (This would be an essentially fair tax since state support for university education involves a vast transfer of funds from the poorer majority whose children are not university educated.)

Heavy child taxes. No government could switch from child allowances to child taxes without destroying itself. However, it might introduce substantial *bonuses* for first and second children that are then rapidly neutralized or heavily reversed by later children. For example, a third child could reduce the previous child grant to nothing, a fourth turn it into a heavy child tax. (There would have to be exceptions for multiple births.) Such a scheme would greatly benefit the poor, small family, while the cost to the state could be offset by recovering the bonuses from the rich by other forms of taxation – as is now done in Britain, where increased family allowances are taken away from higher earners by reducing the child allowance that can be offset against income tax. What no child tax scheme has ever solved is the problem of the poor, large family. Indeed, because of them there is continual pressure to *increase* child allowances. However, though the idea is anathema to many people, a means test would do much to get round this, so that exceptions could be made for families in hardship.

Several other proposals, such as not awarding public housing on the basis of family size, reducing child allowances all round, stopping maternity grants or paid maternity leave, seem to me to be either highly discriminatory or likely to have little effect.

The real point of these lists is not that they offer easy solutions for curbing population growth. What they really show is that if we want to see growth curbed or even reversed we have to start thinking – and thinking unconventionally. Marriage and breeding patterns and the social controls that we already (unwittingly) exert on them are not inscribed on any sacred tablets. They are not God-given laws. They vary widely in present-day societies and have varied widely in every society's own history. Industrial societies have stringent sanctions against illegitimate children, but in some societies a man would not dream of marrying a girl unless she had borne one or two children and proved to be a suitable mother. In Europe and America the fashion now is to marry young, but a century ago our ancestors in the same countries were deliberately marrying late to curb a birthrate they thought excessive.

Similarly, we will change our own patterns and may have to accept social pressures to force the change. Better contraceptive technology and its wider use may help to curb population growth, but even when we all have only as many children as we want, when we want them, there will be a lot of controlling left to do. The chance that the numbers we collectively want would stop population growth is very small, and anyway our desires are bound to swing up or down in future. A true population policy necessitates continual and perpetual control of numbers.

Are we really so afraid of this idea that we will not or dare not face it openly in a full public discussion? Or might we soon begin to realize that a proper population policy could be the lesser of two evils and that anyway it need not be as intrusive as all that? Once we do realize this, and start talking about it, governments will be able – or will be forced – to back the thinking and research that could produce attractive, or at least neutral controls. The longer we delay the more likely it is either that we will be swamped by our own numbers or that the controls, when they do come, will not be attractive or neutral. The choice is ours.

# 3. TEST-TUBE REPRODUCTION

The best way of looking at test-tube reproduction is to climb a kind of ladder of unnaturalness. As we reach each rung – each still more extraordinary reproductive technique – we can ask why and how it might be used by people, what advantages and problems it could bring, and what moral and other attitudes there are to oppose it. On each rung we can ask: Is this where we draw the line?

The ground the ladder stands on is the ordinary way of sexual reproduction: copulation without the least idea of what the product will be like. For most people this is perfectly satisfactory. At least, it produces babies and for the most part they are normal babies. But for quite a large proportion of all humans it is not. Either they are so subfertile that they cannot reproduce at all, or they may know they have some serious genetic defect that they dare not risk passing on. For them artificial reproduction offers at least a chance of a way out.

Whenever we recoil in horror at a new rung up the ladder, we have to remember that if we refuse to go up any further we condemn some couples – perhaps very few – to childlessness or to the risks of conceiving a genetically abnormal child. In earlier chapters we looked at our right to have as many children as we want. Now we have to ask how far this right goes. If society has the means, however extraordinary, to enable a childless couple to have children, does that couple have the right to expect that the means will be used? In some ways, this is similar to the dilemma raised by artificial organs and by transplants (Chapters 9 and 10). Here we are concerned with extraordinary means of enabling people to give life, there we shall be dealing with extraordinary means of preventing death. But there is a difference.

more accurately than she, the physical, emotional and intellectual qualities of the test-tube progeny.

Last, to add respectability, here is part of the statement issued by a group of seventy-four distinguished American scholars, lawyers, editors and economists formed in 1967 and called the Commission to Study the Organization of Peace (as reported in the *Guardian*, London, 5 September 1967):

The most dangerous threat to the privacy of the human being has to do with genetic control. The report believes it will soon be possible to predict the characteristics of a child before birth and soon after that to 'manipulate these characteristics in such a way as to change the future of the human race in accordance with the preconceptions of those who are in control of the means of effecting these changes'.

Why are these people, and many others like them, plainly scared out of their wits? Why are they so sure that these new biological powers must be used for evil purposes and be controlled by a powerful few – by Them?

Obviously, the reason lies partly in the sheer novelty of some of the reproductive techniques which are becoming possible. Sexual reproduction by intercourse between two people is a tradition as old as man, and one does not introduce the possibility of interference by third parties, or the idea of doing away with normal sex altogether to produce children, without seeming to threaten deeply entrenched values. Talk about new techniques of reproduction appears to cut at our own image of ourselves as human beings and lays us open to fears of depersonalization. From there it is a short step to the fantasies of the state-run breeding factories of the Brave New Worlds.

I believe there is an equally potent but less obvious reason. Nowadays there is hardly an article on advances in genetics or reproductive biology that does not end with a (usually) well-meant warning about the awesome problems these will raise for society. But very rarely are these problems actually stated or analysed; almost invariably the warning is general and vague. And when most readers are not mentally equipped to think the thing through for themselves, because they do not have the facts, it is natural that they should be worried. Also, very often biological progress is compared with the surge of discovery in physics forty years ago which revealed the 'secrets of matter' – and eventually presented us with the possibility of nuclear annihilation at the press of a button. And the implication is that biologists, by revealing the 'secrets of life', will produce a threat of the same kind.

Biological progress *is* raising problems which we ought all to recognize and discuss. But they are not problems which compare with the threats raised by the physical sciences. Apart from biological warfare, there is no biological equivalent of the Bomb, no biological button with which a few can gain power over many. However extraordinary the technique, human reproduction can only be manipulated on an individual level and with our consent and cooperation.

We have, then, to inform ourselves sufficiently to be able to decide what we will cooperate in. Only by doing this can we judge what, of all the things that might be technically possible, is both likely to be applied (because someone needs it) and desirable (despite any dangers to individuals or society). We have a real responsibility here, but it is not one which we can fulfil in a panic.

# INTERLUDE: ON KEEPING CALM

In the next few chapters we head out into the wild lands of human reproduction, the domains of the artificial inseminator, sperm and egg banks, and the genetic manipulator. We shall be dealing with novel and extraordinary techniques that arouse extreme reactions, three examples of which are given below.

First, a woman writing in *The Times* (10 December 1967) after a B.B.C. television programme on the subject of these chapters called, significantly, 'Assault on Man':

I watched with increasing horror and fear while scientists from the United States and Britain discussed and illustrated experiments . . . to try and create different forms of life by combining the cells of different species, i.e., man and mouse, man and monkey, etc. An infinite variety of computations can produce an infinite variety of creatures to fulfil certain functions, such as species of subnormal intelligence to provide willing domestic workers, specially ferocious species to produce invincible armies, and so on . . . Are we to sit quietly by while scientists produce the means to destroy the human species in the process of creating other forms of life? Because surely what they are unleashing will be beyond the control of even the most responsible society.

Second, the *New Scientist*, in a lighter but no less revealing mood, wondering about the role of the future biologist as a none-too-benevolent fairy godmother (36 [1967], p. 745):

In future, by feeding into the fertilization process the necessary hereditary influences, we might be able to breed whatever type of person we desire. Worker-people, soldier-people, thinker-people and all-sorts-of people, might be ordered in advance according to the state of the nation and the requirements of the market. Sitting in his sperm store in his white lab coat, the birth controlling biologist may not be as fetching as the pretty lady in her diadem, tutu and tights, but he'll be able to predestine

With reproductive techniques the end result is another human being. We not only have to consider his welfare, we also have to bear in mind the welfare of the unwanted or orphaned child who might have been adopted by the childless couple if this ladder had not been made.

So now let us climb the ladder and pause every now and then to see whether we like where we have got to.

## Rung 1. The infertility clinic

This is a staging post for sorting out those who in practice need to climb the ladder. Leaving out genetic reasons for the moment (see Chapters 4 and 5), infertility is the most likely reason why they should want to. Though very few couples are completely sterile, about 10 to 12 per cent of all marriages are so subfertile that even fifteen years of trying does not produce a baby. For Britain this means about 40,000 new marriages a year. Tradition has always blamed the woman, but it is now known that around 10 per cent of infertility is due entirely to the husband and in another 40 per cent of infertile marriages it is largely due to him.

After a few years of trying, a childless couple may ask an infertility specialist for advice at a clinic. One of the first things he may discover is that sexual intercourse itself is the trouble. Many infertility clinics report that one in every twenty women coming to them is a virgin; that with many others sexual technique is so poor that the husband's sperm has little chance of reaching the cervical opening at the top of the vagina, or even the vagina itself; and some wives say they have intercourse only three or four times a year.

If intercourse is normal, tests start. First, the sperms are usually tested by seeing how many survive a few hours after intercourse. If the number is abnormally low (less than about 10 million instead of the normal 200 to 300 million) there may be another sperm count after masturbation. If this count is very low, sperm *production* could still be normal, but an ulcer or a 'growth' may be blocking transmission.

For women tests are more varied. If the husband's sperm is normal the wife's cervical mucus may be deficient, missing or infected, or the Fallopian tubes may be blocked, or the woman may not produce normal eggs (ova). The latter can often be checked by a simple test, while hormone treatment – for example with F.S.H. (follicle stimulating hormone) or Clomiphene – can now often induce normal ovulation, though with considerable risks of multiple births.

After all these tests and treatments, and also psychotherapy for impotent men or clinically frigid women, more than 50 per cent of younger couples and about a third of older ones (women over thirty) will be likely to conceive normally. They can go back to ground level. For the rest, the choice is childlessness, adoption or the next rungs of the ladder.

## Rung 2. Artificial insemination by the husband (A.I.H.)

Despite what many people think, artificial insemination by the husband is far less common than artificial insemination by a donor (A.I.D.). The commonest reasons for it are malformed male organs, impotence, very low virility where intercourse is rare, or premature ejaculation. In other words, there are usually quite serious psycho-sexual disturbances in the marriage relationship.

A couple can carry out A.I.H. themselves. The husband masturbates and a syringe is used to introduce the sperm directly into the cervix. Usually an artificial-insemination specialist does it, because he can more accurately gauge the wife's most fertile period: a few hours either side of ovulation can make a big difference to the chance of success.

The other main reason for A.I.H. is gross male subfertility. The idea is to place the few, weakly-moving sperm directly in the cervix and cervical mucus to give them a shorter and less acid journey to the egg. But results are remarkably poor: success rates of 20 per cent are exceptionally high, probably because a low sperm count goes with other sperm deficiencies.

## Rung 3. Sperm bank A.I.H.

Sperm can be frozen and stored indefinitely. This is usually done by mixing the specimen in a flask with a protective agent like glycerol or egg yolk and cooling slowly to $-196°C$, the temperature of liquid nitrogen. It is then stored in nitrogen in a thermos flask. An obvious use for a sperm bank is to store up several ejaculations from a grossly subfertile husband – perhaps over weeks or months – to bring the total up to normal levels. Thawed out, they can then be used for normal A.I.H. Another use is for inseminating a wife several times in the couple of days around ovulation to make sure that at least one insemination is at the best time. Unfortunately, results are very discouraging so far, because 'poor' sperm does not seem to survive freezing and thawing well.

*

Before going any further, we must consider what objections and problems we have passed by on our way up the first three rungs.

With the first – the infertility clinic – no one seems to raise any significant objections. Some people ask why society should spend its resources on helping the infertile to have children when there is a population explosion occurring, but these people are very rarely infertile themselves.

As for problems, cost can be very important. Hormone cures for infertility can run from about five hundred pounds to an exceptional five thousand pounds for a really difficult male. Though costs will almost certainly drop, in a state health service this does raise a serious question. At what level of cost does the doctor (and society) suggest adoption? How *does* one assess the value of a long-sought child?

But for me the most serious problem comes from those pathetic 'infertile' virgins, and the fact that many infertile couples are too shamed by the stigma of their affliction, or too shy, to go to a clinic at all. (When will sex *really* become a topic for open discussion?)

With the second rung – A.I.H. – real controversy begins. The

Catholic Church totally bans it, and every technique from here on, as an unwarranted interference with the natural processes of sex and procreation. According to Catholic teaching, marriage does not give a husband and wife an absolute right to conceive children, but merely to have sexual intercourse in a way that might lead to conception. In 1949 Pope Pius XII rejected artificial insemination absolutely because, among other things, it would place man on a level with beasts and would 'convert the domestic hearth and sanctuary of the family into nothing more than a biological laboratory'.* Despite this, some Catholics do not object to 'assisted' insemination, where after intercourse a syringe is used to project semen into the cervix. The Jewish Orthodox Church and the Lutherans also prohibit all forms of artificial insemination. By and large, all other Christian Churches, though not exactly welcoming it with open arms, are not now opposed in principle.

The legal position is astonishing. By mid-1967 only two communities in the world had any laws concerning artificial insemination in any form. There was a New York City health regulation governing the use of donors and a (May 1967) law in Oklahoma which does go a long way towards clarifying the tangled legal issues of A.I.D. (see Rung 4). With A.I.H. this is not too important, since the only likely reasons for a court case are where an impotent husband agrees to A.I.H. and then he or his wife seeks an annulment of marriage on the grounds of nonconsummation; or where a husband or wife persuades a doctor to carry out an A.I.H. without the other's consent. With the first, most lawyers now agree that for the child's sake an agreement to A.I.H. should be counted as consummation. In the second, legislation is hardly relevant, since virtually every artificial-insemination practitioner now insists on both marriage partners signing a consent form.

Perhaps the most important A.I.H. topic that needs airing is the doctor's right to turn applicants away. This arises because most A.I.H. husbands have fairly serious psycho-sexual diffi-

* G. Kelly, 'Teachings of Pope Pius XII on Artificial Insemination', *Linacre Quarterly*, 23 (1956), pp. 5–17.

culties and nearly all A.I.H. practitioners, because they feel very responsible for the child they are helping to produce, also feel they must act as judges of the marriage in case it goes badly wrong and threatens the security of the child. So in many cases they refuse A.I.H. This is a unique, almost god-like position for doctors to put themselves in, yet it is strongly defended by most A.I.H. practitioners. Though there are many medical precedents for this kind of attitude – elective rather than emergency operations are often very much the doctor's choice – none are as extreme. If the A.I.H. doctors were absolutely sure they had the psychiatric sophistication to make these judgements properly, it would be arguable that their case was a proper one. But obviously few do, and certainly they do not always call in expert psychiatric advice. Neither would their attitude matter much if there were a vast pool of other artificial-insemination doctors for a refused couple to turn to. But there isn't: such doctors are very thin on the ground. So the couple may well have to remain childless, or adopt (which brings us back full circle). Of course, there is another alternative: psychiatric treatment for the couple.

If we, as a society, believe that people have the right to assistance in conceiving children, at least up to this point on the ladder, it is clearly intolerable that doctors should act in this way. But as with abortion and contraception, it is not easy to insist that any doctor must do something that goes against his conscience. It may be that just as abortion technicians might solve the abortion problem, so artificial-insemination technicians, or a structure which allows the 'liberal' doctors to do much more, could solve this one. But we cannot avoid the insistent question: is the 'right' of a childless though disturbed couple to have children, when that is possible, stronger or weaker than the doctor's 'right' to decide they are not fit to have them?

The third rung – A.I.H. with sperm banks – raises no extra problems. The bank is merely a tool for making A.I.H. marginally more effective.

## Rung 4. Artificial insemination by a donor (A.I.D.)

The first recorded case of A.I.D. comes from an Arabian manu-
script of 1322. A horse-breeder left a wad of wool in a mare
overnight, took it the next night to the nostrils of a prize stallion,
caught the ejaculated seed on the wool, and introduced it in his
mare – which duly foaled. Since then A.I.D. has become an
essential part of modern animal husbandry: in Denmark, for
instance, 98 per cent of all calves are sired by A.I.D. from a few
bulls.

The first A.I.D.s with humans were done in the 1890s by an
American doctor, Robert Dickinson, in great secrecy. Since then
their number has increased slowly but steadily until today the
usual (very rough) estimates are 5,000 to 10,000 A.I.D. children
a year in the U.S.A., and perhaps 1,000 in Europe, a few hundred
of them being British.

Artificial-insemination doctors broadly agree that the most
usual valid justification for A.I.D. is in the case of a psychologically and physically normal woman married to a male who is
similarly normal except that he produces no sperm at all (azoospermia). A.I.D. may also be used if the sperm count is very low
and the couple have tried for a long time; and where there is a
grave risk of genetic defect. Though azoospermia accounts for
only about 5 per cent of all 'sterile' marriages, or a mere 0·5
per cent of all marriages, the potential demand is still surprisingly
high: about 2,000 new marriages a year in the U.K., and 8,000
in the U.S.A. Of course, many of these couples will prefer
childlessness or adoption, though as A.I.D. becomes more
usual probably fewer will. Yet already several surveys have
shown that more than 50 per cent of infertile couples would
choose A.I.D. rather than adoption, and that it produces fewer
problems afterwards.*

The main advantage of A.I.D. over adoption is that the wife
has the full experience of motherhood. Many husbands find they

* For references to these see the chapter 'Artificial Insemination' by
S. J. Behrman in Behrman and Kistner, ed., *Progress in Infertility* (Little,
Brown, Boston, 1967).

can share in this experience too, just as though they were the true genetic father. The process can also be far more simple and *private* than adoption: a few visits to the A.I.D. doctor plus an apparently normal birth, rather than the long scrutiny by the adoption agency and the sudden arrival of a six-week-old in the house to arouse the neighbours' curiosity. Adopting parents are also often afraid of the natural mother, however many legal bits of paper they have. The anxiety of a pregnant woman who thinks she may miscarry is nothing to the terror of the adoptive woman waiting to see if the natural mother will reclaim her child before the adoption is finalized. Also, with an A.I.D. brood the donor is matched to the husband's obvious characteristics, the mother is the same, so the children all look 'natural'.

Briefly, the technique is this. Having assessed a couple as being suitable, preferably after a full psychiatric evaluation (still rare), the artificial-insemination doctor selects a donor from his private donor pool. Standards here vary widely. Some doctors insist on married men with children and free donation, others use medical students for a fee (usually five pounds, or in the U.S., where this is more common, fifteen to thirty-five dollars). All insist they do a thorough health check on donors and a genetic search in case of hidden defects that might appear later. They also try to select donors who match the obvious physical characteristics of the husband – build, eye- and hair-colour, and so on – as closely as possible. 'Intelligence' matching, or even raising, is fairly usual.

The wife is ready for insemination at her next mid-cycle ovulation, and goes to the doctor's surgery. The donor, warned in advance, produces his sperm specimen by masturbation. Ideally, this must be done not more than thirty minutes before the wife is inseminated. To keep his identity secret from the couple, the donor does *not* deliver the specimen bottle himself to the surgery: an employee of the doctor (sometimes a taxi-driver) collects it.

All artificial-insemination doctors insist on strict donor anonymity and often have elaborate coded filing systems in case a wife gets too curious. Some, however, will occasionally allow a close friend of the couple to donate, or a male relative of the husband's. The insemination itself is a simple business: about a saltspoonful of donor semen is syringed into the wife's cervix, she lies down for twenty minutes, and it is done. Some doctors carry out another insemination the next day (using the same donor again) to increase the chance of fertilization. Many also advise the couple to have intercourse within a few hours of the insemination, or even mix some of the husband's sperm with the donor's. The idea is to let the husband believe that the child might be his; and, in case of legislation, to make it difficult to prove A.I.D.

All A.I.D. is private, fee-paying medicine. This is partly because it is considered 'luxury' elective medicine, partly because no health service has yet seen its way to shouldering the responsibility to the donor and the child that private artificial-insemination doctors feel they bear. Costs vary widely. One American with a large practice (six inseminations a day) charges $75 per patient per month (for three attempts) and estimates that 70 per cent of his clients are pregnant within four months. If they are not he makes no further charge, and gives up the case after about six to nine months as hopeless. So the maximum charge is around $300. A leading London practitioner told me he charges £80 to £100 per course of treatment, even though this very often involves only a single insemination. Other London practitioners would not disclose their fees, but all said that £100 was fantastically high.

As with A.I.H., doctors insist on the right to refuse. How often they exercise this right varies enormously. One London doctor I talked to turns down three couples for each one he accepts; another has about 120 applicants a year and accepts roughly 100; another (a woman) who does about 50 to 60 A.I.D.s a year says she has turned away only 10 applicants in five years, mainly on grounds of social inadequacy. All say they have to do this, since they cannot go against their own consciences, and all but one impressed me by their responsibility. The exception boasted that he had a sharp eye for the frivolous request – the couples who 'come looking for babies like shopping for shirts'. He had recently seen a stage couple who, he agreed, had thought hard about A.I.D., knew just why they were asking for it, and were 'charming and responsible' people. But he turned them down 'because we all know what stage marriages are like'.

There are three main A.I.D. problem areas: religious or moral, legal and psychological. The main religio-moral concern is whether A.I.D. is compatible with the time-honoured traditions of the sanctity of marriage. In other words, is it adultery? Many moralists sincerely believe it is. For a wife to receive the seed of a third party, even though the motives and consequences for the marriage may be wholly good, goes against the essential nature of marriage and should be forbidden.

One of the fiercest statements of this view (outside Rome) was made by the Archbishop of Canterbury's Commission on A.I.D. in 1948:

Artificial insemination with donated semen involves a breach of the marriage. It violates the exclusive union set up between husband and wife. It defrauds the child begotten, and deceives both his putative kinsmen and society at large. For both donor and recipient the sexual act loses its personal character and becomes a mere transaction. For the child there must always be the risk of disclosure, deliberate or unintended, of the circumstances of his conception. We therefore judge artificial insemination with donated semen to be wrong in principle and contrary to Christian standards.

Ten years later a Church of England working party, preparing evidence for a large inquiry into artificial insemination conducted by the Home Office,* adopted this statement in its conclusion and added: 'We hold A.I.D. to be wrong on theological, moral and social grounds.' A further ten years have

* *Report of the Departmental Committee on Human Artificial Insemination* (Command 1105, July 1960), (H.M. Stationery Office, London). This is an exhaustive (90-odd page) examination of all aspects of artificial insemination which makes irresistible reading, especially if one likes picking deep holes in reasoned arguments. The Church of England evidence is published as *Artificial Insemination by Donor* (Church Information Office, London, 1959). The first half, by the Archbishop of Canterbury, is a classic instance of the Church fighting innovation by ignoring all evidence but its own. But to be fair, it was written when the general climate of opinion on A.I.D. was far more intolerant than it is now.

softened this attitude considerably, and the majority of Protestant theologians and moralists now point to the profound differences in motive and practice between adultery and A.I.D. With the first the aims are sexual pleasure and an emotional involvement which rejects the husband and usually tries to avoid a child. With A.I.D. the married couple consider and consent after hard thought, and want a child to rear inside the marriage, while the wife never meets the donor or even knows his name.

The law, too, has softened its attitude recently. In the 1920s there were splendid judicial pronouncements on the marital sin of A.I.D. The most deliciously concise came from Lord Dunedin, in Britain: 'Fecundation *ab extra* is, I doubt not, adultery.' In a 1954 divorce case an Illinois court ruled that artificial insemination with or without the consent of the husband was immoral, the wife an adulteress, and the child a bastard. Since then common sense has crept in. In England and Scotland, for example, case law – it is all case law – probably depends on a decision by Lord Wheatley in 1958 that A.I.D. does not count as sexual intercourse and adultery demands that both parties (oh, how fine the mills of the law grind) must 'engage in the sexual act at the same time'.

The second main target for moral and legal concern is the status of the child. Is he legitimate? Is he legal heir? Who is his father? Who has custody after a divorce? Until recently there was violent opposition to A.I.D. on these grounds, with much fearful talk about the age-old traditions of the inheritance of property, titles and so on. Now opinion is slowly swinging into line with the simple statement of one woman who had an A.I.D. child: 'My husband is our baby's father. Fatherhood is more than genes and chromosomes.' In other words, morality and law are beginning to grant the A.I.D. child at least the status of an adopted child, and, many would insist, that of a true child. This would prevent the recurrence of those extraordinary cases where a husband has been deprived by a court of all rights to 'his' child after divorce proceedings, even though he gave full consent to A.I.D. and protested vehemently that he loved and cherished the child as if it was his own. Any dispute over the legal state of the child can of course be obviated if the couple legally adopt it. Unfortunately, most parents refuse to do this, as it defeats one of their basic concerns – keeping the A.I.D. secret.

A third main legal concern has been over A.I.D. without the husband's consent. Is it, for example, grounds for divorce? The majority answer is 'yes'. Though thousands of man-hours must have been spent reaching this conclusion, the fact is that no doctor would dream of doing an A.I.D. without a full consent form – including a waiver of rights in the child from the donor.

Standing out from all the other, more detailed legal and moral problems of A.I.D. are two crucial general points.

First, the law has been extremely slow in tackling the challenge of A.I.D. Officially, A.I.D. does not exist. It is a blank page in the statute books, and case-law opinion is remarkably varied. Meanwhile, many artificial-insemination doctors insist that the law's failure to act has been one of the major deterrents to the widespread use of A.I.D.

The second point is the astonishing distance attitudes have swung in a short – but not all that short – time. As the American artificial-insemination practitioner Dr Sophia Kleegman has put it, 'Any change in custom or practice in this emotionally charged area has always elicited a response from established custom and law of horrified negation at first; then negation without horror, then slow and gradual curiosity, study, evaluation, and finally a very slow, but steady acceptance.'* Horror is no friend to truth, so in the early stages even the most 'responsible' commentators let prejudice conquer fact. To give just one example, the big Home Office report already mentioned hits hard at the dubious morals and lack of responsibility of donors. It is also worried that these characteristics might be passed to the child. What evidence there is suggests that unpaid volunteer donors are highly moral and responsible, and that their main motivation is, simply, to help in cases of involuntary childlessness. As for inheriting immorality through the genes, that is a rather novel concept in genetics. One cannot help wondering how much of the

* S. J. Kleegman and S. A. Kaufman, *Infertility in Women* (F. A. Davis & Co, Philadelphia, 1966).

moral fervour is *really* about masturbation. The donor cannot produce his donation in any other way, and for many people this is enough to make his gift repellent, instead of being admirable, like blood donated in a nice sterile syringe. Reactions to the first female donors, where this does not apply, will be interesting to watch.

The most important A.I.D. problems are surely not legal or moral but broadly psychological. How does it affect the people most intimately concerned – a husband, a wife, a child? Despite many thousands of A.I.D.s, the most honest answer is that no one really knows. The reason is interesting. Nearly all A.I.D. doctors believe that the couples they help should not be kept under observation. This would remind them that their child was by A.I.D., and without reminders they can usually ignore the fact that it was. As one doctor told me, 'Never sacrifice human happiness for scientific results.'

What evidence there is seems to oppose the worst fears that A.I.D. is a psychological disaster. In fact, if it is carefully handled, the psychological outcome can be remarkably good. For example, S. J. Behrman, an American A.I.D. practitioner, reported in 1966 that in 393 A.I.D. pregnancies there were only 2 cases of 'emotional disturbance', and in 1967 claimed* that of 800 A.I.D. couples only 1 had had a divorce. Since one in four American marriages ends in divorce this says a lot either for his psychological competence or for the maturity and stability of marriage partners who choose A.I.D.

On the other hand, there can be dangers. An American psychiatrist, Dr Bernard Rubin, has reported† that out of 43 women who had A.I.D. all had actively to suppress fantasies about the donor, while a third were preoccupied with his looks and personality. A French gynaecologist‡ tells how he and many colleagues have given up A.I.D. work because of disastrous failures. In one case a young wife learned the name of the donor and abandoned her husband to join him; in another, a wife bore

---

* See footnote p. 86.

† 'Psychological Aspects of Human Artificial Insemination', *Archives of General Psychiatry*, 13 (1965), pp. 121–32.

‡ Jean d'Alsace, *Gynécologie pratique*, 13 (1962), pp. 707–10.

a strikingly beautiful child by A.I.D., started to dislike her husband and pestered the gynaecologist for the name of the donor, whom she wanted to marry. But most artificial-insemination doctors insist that bad cases like this must be very rare and are probably due to poor psychological handling and screening before A.I.D. If the husband completely accepts that he is sterile; if he sincerely wants his wife to have a child rather than adopt one; if neither husband nor wife vacillates; if the husband wants the child as 'his'; if the wife has no qualms about her future attitude to the child or her husband; and, above all, if the wife realizes that a husband's agreement to A.I.D. is a profoundly important gift to her – then A.I.D. can give excellent psychological results. But this is a long and difficult list for any doctor to go through and any couple to satisfy. And clearly it suggests several questions that need answering by the professions concerned. Should there be much more psychiatric study of A.I.D.? Should psychiatric assessment of couples be compulsory? Or should medicine and society accept some risk of psychological disturbance if the couple concerned are ready to take it as part of the price of having a much-wanted child? After all, many children conceived in bed have psychologically disturbed parents and have to live through divorce and every kind of hideous insecurity. All too many, as we saw in the first two chapters, start life unwanted. This, at least, is presumably spared the A.I.D. child. Wanted he must be, or why did the parents bother? If *natural parentage* is a right, then A.I.D., as being the nearest approach for these couples, is difficult to condemn. If *rearing a child* is a privilege, then *all* parents should presumably be screened as carefully as adoptive ones are. Nevertheless, until we know a great deal more than now about what makes a 'good' environment for child-rearing it would be rash to assume that A.I.D. necessarily produces an increase in happy, 'well-adjusted' children.

*Rung 5. Sperm bank A.I.D.*

Sperm bank A.I.D. is now fairly common. When one thinks of the logistic problems of A.I.D. the reason is obvious. Without a bank the donor has to provide his specimen to order, probably on two successive days each month, perhaps for several months on end; and it has to get from donor to doctor, in secrecy, within thirty minutes. This is a strain, so few donors will donate for more than about half a dozen patients. This means that A.I.D. doctors are for ever having to search for willing donors and screen them medically and genetically. Some admit they do not do this as well as they would like. There is also the husband-matching problem. With a donor shortage it is not always possible to get a healthy, high I.Q., genetically cleared donor who also matches the husband's physical characteristics.

Large central sperm banks would help enormously. They could store a wide variety of 'sperm types', including unusual ones (as regards race, blood groups, height and so on). All donors could be rigorously screened – as blood is, in the best blood banks. There could also be a high 'secrecy factor'. The artificial-in-semination doctor would ask the bank for so many cubic centi-metres of semen of such-and-such donor characteristics, and the only clue to the donor's identity would be a tube with a serial number on it. Of course, there would have to be fool-proof secrecy codes at the bank, but these are easy to devise. Such banks, though on a less grandiose scale, have been established. The first two, set up in 1964, were in Iowa City and Tokyo.

With sperm bank A.I.D. we hit some unusual biological prob-lems. One is the risk of 'innocent incest' – the chance that two A.I.D. offspring of the same unknown donor might marry. This is true of all A.I.D., but with fresh sperm it is minimal. For instance, if there were 2,000 A.I.D. births in Britain each year (roughly ten times more than now) and each donor is used 5 times (which is roughly how often they are used now), an un-witting incestuous marriage would occur only once every 50 to 100 years. This is said to be less than the risk of an early-separated natural brother and sister marrying each other. But if a national sperm bank led to a heavier demand on some donors, this risk could go up sharply. So there might have to be legal limits on how often each donor is used, or even a law that the central donor register must be consulted secretly before every marriage.

There are other genetic dangers, too. The first sperm-bank child was born in the U.S.A. in 1953 and was perfectly normal. Since then numbers have slowly swelled and storage times increased up to three years or so, and there have been no genetic disasters. But these are early days: what few reports there have been tend to be about a mere ten or twelve children. The fact is, no one knows. Neither does anyone know about another, possibly much more dramatic genetic consequence of sperm banks. A sperm bank is a kind of evolutionary forcing ground in miniature. There is a very intense selection of sperm for freezability. Sperm samples from different donors vary enormously in their ability to withstand the freeze-store-thaw process. There is also intense selection within any one sperm sample from whatever donor. Since 'freezability' must in some way be a genetic trait it is extremely likely that it is connected with other genetic traits that matter – like disease resistance, stature, intelligence and so on. So sperm with these traits will tend to survive and the traits to be passed on. At present all one can say about this is that there will be a lot of controversy – perhaps on the scale of that on the genetic effects of radiation.

Another serious problem is the possibility that central banks *might* be misused. As long as sperm banks are a matter of a thermos flask or two in a doctor's surgery, misuse is unlikely. Doctors are on the whole fairly ethical characters. But with a central bank a government could conceivably take charge and start dishing out their own choice of sperm. I cannot see any present government doing this, or see what any government would gain by it. But once there is central secrecy, the possibility is there. There is also the danger that shady commercial operators may get in on the sperm-bank business. Some American undertakers are said to offer sperm storage as part of the burial contract. This may be a harmless sop to the ego of the dying as long as the sperm is not used, but it could spread to the peddling of promises of high-grade children for the living. We shall go into this further in Chapter 4, where we look at the old eugenic dream that man can improve himself genetically – a dream made possible only by sperm bank A.I.D. But in the meantime one

might note that in 1937 a move was made by a European gynaecologist to set up a Seminological Bureau to collect sperm
specimens from as many contemporary celebrities as possible.*
The idea was partly to start a rather curious museum, a kind of
sperm cellar whose value would mature with time, but also to
make a handsome profit by inseminating the gullible.

This brings us to one of the most extraordinary consequences of
sperm banks: the fact that they allow the traditional space-time
lock of sexual reproduction to be broken. With them, in principle,
a woman can conceive by any sperm sample from any man even
though he is long dead.

## Rung 6. 'Space-time' sperm banks

There are several reasons why women (and their husbands)
might want to use sperm banks to break the traditional need to
be together for conception. Most of them involve A.I.H. rather
than A.I.D., but there are good reasons why they should come
so far up our ladder of unnaturalness.

*Insurance deposit account.* The most likely reason, a husband
deposits his sperm in a bank before he is sterilized for contraceptive reasons. He can then draw them out if the sterilization is
irreversible and he and his wife decide later to have children, or
he remarries. *A.I.H.*

*Mutation protection.* Some biologists have argued that all males
might store their sperm at puberty in lead-lined radiation-proof
sperm banks to protect them from mutation damage, which acts
fairly steadily through life. They could then draw on their banks
later, when married, to have children. This is unlikely to be
widely practised, if at all. The psycho-sexual distortions of the
technique would surely be greater than the biological risk that
is being avoided. Who bothers to have babies quickly because of
mutations?† Men working at sustained high-radiation levels,

---

*For more details see the *Eugenics Review*, 48 (1957), pp. 209–10.
†An odd thought. At the height of the furore over the genetic damage

including astronauts and supersonic transport crews, are the most likely candidates. *A.I.H.*

*Germinal choice.* As we shall see in Chapter 4, some normal, fertile couples may in future decide for eugenic reasons to have a child or two using a 'high-quality' donor who has deposited his sperm in a bank, and may be long dead. *A.I.D.*

In the above three cases the husband and wife presumably decide jointly whether and when the technique will be used. In the next two cases they do not.

*The reverse chastity belt.* A couple may wish very much to conceive a child when the sailor, businessman or astronaut husband is away on a long trip. The trip would have to be long (two years?) to make it worth bothering about. Though the decision may be a joint one, it is up to the woman alone just when she decides to conceive. *A.I.H.*

*The widow's last link.* Sperm banks allow a widow to bear children by her dead husband. Because of the emotional power of this possibility, it could well be the strongest reason for wanting to use space-time sperm banks. *A.I.H.*

The main danger of this Rung 6 has already been hinted at. In the last two uses, reproductive function is partly or completely separated from a *continuing* sexual relationship. At present when a woman becomes pregnant we assume she is living with or loving a man. Here she is in a completely autonomous position over her child-bearing: she has a child by her husband when *she* wants to, not when *they* want to. Furthermore, such children would be peculiarly the mother's property. The absent (or dead) husband might have no part in its rearing. Emotionally if not socially, it would be in a position similar to illegitimate infants whose fathers

---

from bomb-testing fall-out, when masses (including myself) marched with protesting banners, how many people protested about the genetic effects of contraception? Delaying the birth of a baby by four months exposes the parents to about as much extra radiation as did fall-out – i.e. roughly 1 per cent of the natural background radiation accumulated over 30 years.

are unknown or unadmitted. Largely for these reasons, all artificial-insemination practitioners but one to whom I have talked are adamant that they would never take on a widow – the only case of the five possibilities listed which they are ever asked about. The exception, a leading American, said he would help a widow with a few children, but not if she had none or if she had married again.

## Rung 7. Egg grafts

So far in this chapter only male sperm has been mentioned, and in fact only male sperm has ever been given, stored and used. There has been no transfer of female germ cells, and therefore no 'cure' for female sterility. Now we move to a new level: reverse insemination.

Some women have no ovaries; others fail to ovulate despite hormone treatment; others may have a genetic affliction. For any of these an egg graft would offer hope of a normal child. Briefly, a donor woman supplies a ripe egg when she ovulates, or she may be dosed with hormones to make her 'super-ovulate' several eggs. The egg is then implanted in one of the Fallopian tubes of the wife, where it is fertilized by the husband in the usual way. At first sight this seems to be the 'mirror image' of sperm A.I.D. Actually, it should be far easier psychologically. For one thing, conception is by normal coitus. But above all, though the wife is not the genetic parent of the child, because she carries and bears it and because the husband *is* the true father, husband and wife share parenthood more equally than with conventional A.I.D. It may even be that for women genetic parenthood is less important than it is for men. A woman's genetic contribution is 'invisible', while her environmental contribution lasts nine months and culminates in the emotional and physical upheaval of birth. For men, their ejaculation is their sole physical contribution to the child.

On the other hand, the technique is much more demanding. The donor has to have her menstrual cycle synchronized with the wife's by hormones, a process that may take a few months.

Though removing her egg is comparatively simple – it can literally be washed out of her Fallopian tube with a pipette – implanting an ovum in the wife requires at present a minor surgical operation (a telescopic tube, or culdoscope, carrying the egg is inserted through the cervical opening and up a Fallopian tube). The wife would surely face this, but getting donors may be far more difficult, especially since they may have to cooperate for much longer than male donors, as success-rates will probably be much lower. Possibly the only women who would agree to donate would be sisters or close friends of the wife, which would of course rule out donor anonymity.

Many artificial-insemination doctors are equipped to use this technique and have been asked to do so by women patients, though as far as I know it has not yet been used.

I can see only one problem not already covered by sperm A.I.D., but it is an absolutely crucial one. It is the same kind of problem that may prevent all those brave dreams of genetic surgery* ever becoming fact. It is, simply, that while fertilization starts with millions of sperm it starts with only one egg; and with all the handling it gets during the graft, the delicate egg may be damaged. If the damage was lethal this would matter less than if the egg survived but with, perhaps, a chromosome defect or two. Unfortunately, there would be no way of knowing there was damage. One way round this is to implant several eggs, but this brings the risk of triplets, or more. Given ways of detecting chromosome faults in the foetus (see Chapter 6) there could be an abortion, but for a woman with no ovaries this would be doubly tragic. To say the least, the first egg grafts will be a testing time.

## Rung 8. Test-tube fertilization

Some women produce perfectly good eggs but have incurably blocked Fallopian tubes, so that male sperm has no chance of getting to the egg to fertilize it. A solution would be to remove some eggs from the ovary, fertilize them with sperm outside the woman's body in a 'test-tube', grow the fertilized eggs for a day

* See p. 153.

or two, and then implant one in the woman's uterus, where it can develop normally. Since the chance of a 'blocked-tube wife' marrying an azoospermic husband is infinitesimal, the sperm would be the husband's. This is therefore an A.I.H. technique: both husband and wife are the true parents of the child.

Test-tube fertilization is not yet ready for the doctor's surgery but might well be soon. In 1966 Dr Robert Edwards of Cambridge University, England, showed for the first time how eggs extracted from human ovaries (removed for therapeutic reasons) could be cultured in the test-tube so that large numbers developed into ripe eggs. Then in February 1969 came the news that he and his team had at last managed to fertilize such eggs with human sperm so that they began to develop.*

By the end of 1970 Edwards, and his gynaecologist colleague Patrick Steptoe, had achieved worldwide notoriety with their premature announcement that they planned to use their technique for the first time on a woman volunteer and had also made great strides in perfecting their methods. In November 1970 they reported successful fertilization of 53 eggs out of 182 human ova, of which 40 had started to divide. In a few cases the eggs reached the blastocyst stage with 60 to 100 cells and were in a 'healthy condition', so that they could have been implanted in a woman's uterus with a fair chance of growing to term. But at the time of writing no such implantations have been made, largely because there is no way yet of ensuring that the test-tube blastocysts are free of chromosome defects, particularly of the mongoloid kind.

From here it is not a long step to clinical use. When this is possible, the eggs would not come from therapeutic operations but from the wife's own ovaries. They cannot be flushed out since the Fallopian tubes are blocked, so they have to come from the ovary itself. This means making a small incision through the wall of the vagina to reach the ovaries and snipping off about a dozen egg follicles, a relatively minor operation.

There are some new problems here. Test-tube fertilization not only involves growing human eggs outside a woman's body – a

* R. G. Edwards, *et al.*, *Nature*, 221 (1969), pp. 632–5.

major break with tradition – it also means deliberately letting some die in the test-tube. This is necessary because the technique is, and may always be, inefficient. To guarantee one successful implant the doctor may have to fertilize, say, ten eggs, may succeed with five of them, and so has to dispose of four viable fertilized eggs. Because of this many people – and not just Catholics – are outraged. Though they are not in the least upset by contraceptives (such as intra-uterine devices) which work by killing fertilized eggs at a far later stage of development, the idea of the death of a micro-embryo in the test-tube seems to them particularly appalling. Is their attitude justifiable? Or is it mere unreason, based on the perhaps crucial difference that the second kind of death is *visible*?

These questions demand clear answers, because before any couple can profit from test-tube fertilization many scientists will have to grow a large number of human fertilized eggs and cut them up. Only by these micro-autopsies can they hope to perfect techniques that do not damage the eggs – and thus avoid the risk of defective babies, or the multiple-birth problem we met on the last rung.

The other serious problem raised by this work concerns the risk that a foetus grown from a test-tube conception will be damaged in some way. It is not yet possible to check that the implanted blastocyst is free of damage and it will probably be a very long time, if ever, before a 'no-damage guarantee' can be given. There is therefore a strong argument for insisting that the technique should not be used on humans until it has been safety-tested by extensive animal trials. Another approach, favoured by Edwards and Steptoe, is to proceed with human experiments but keep a constant check on the developing foetus (by techniques described in Chapter 6) and recommend abortion if genetic abnormalities appear. It may seem somewhat callous to contemplate abortion for women desperately longing for a child, and there is a real chance that once pregnant a woman may refuse abortion whatever the abnormality. Yet this approach may be the only way of granting the chance of child-bearing to many women and is not so different in its calculating acceptance of possible genetic damage as the way all normal couples conceive children (see Chapter 5).

## Rung 9. Egg banks and 'embryo' banks

Once eggs can be grafted, few experts doubt that they will very soon be stored and banked as sperms are now. This seems to raise no new problems.

However, once eggs can be fertilized in the test-tube it should also technically be feasible to store fertilized eggs or micro-embryos (blastocysts) that could be thawed out and implanted in a womb to develop. As we shall see in Chapter 6, it is highly unlikely that these stored 'embryos' will be recognizably human. There is no question of Brave New World banks of bottled foetuses to be picked off the shelf like so much shopping. Instead, they will be cell-masses containing around fifty to a hundred cells and therefore hardly visible to the naked eye. Yet, surely, we ought to be very careful before we allow this to be done. Although unfertilized eggs and sperm are in a sense potential people, a fertilized egg or embryo is a genetically uniquely defined person. The chance combination of fertilization has taken place – and produced a unique blueprint for development. While anyone who argues for post-fertilization contraception, or abortion, should not make too much of this fact – and recognizes more powerful factors weighing against it – the question of *storing* such potential people does seem to be altogether different, and abhorrent. People are born, people die, people are killed; but 'people' have not yet had their life cycles suspended in a deep freeze.

## Rung 10. Host mothers

The host mother could be an animal or a woman. The animal host technique is called hetero ova transplantation. It has been done scores of times with animals, as it is an extremely cheap way of transporting high-grade genes around. For example, top-quality British lambs have been born to run-of-the-mill ewes in

America and South Africa after they were conceived in British ewes, flushed out as micro-embryos, and transported by air implanted inside host rabbits. One small rabbit can carry several embryos, so there is no need to pay freight charges for moving large numbers of pregnant sheep. At its destination the rabbit is killed, and the embryos are removed and then implanted in ordinary ewes, where they develop normally. This far-fetched technique could perhaps one day help women with a fault that prevents the normal attachment of the blastocyst to the uterus wall. The embryo could be conceived by wife and husband, hosted for a week or two by an animal, and then implanted in the wife when it had a better chance of survival there.

The human-host technique stretches this idea even further. Where a woman has no uterus but is otherwise reproductively normal, she and her husband could produce a fertilized egg in the usual way. The egg would then be transplanted to a voluntary host mother who carries it to term, bears the child and hands it back to the couple. In other words, the host mother is used as an incubator.

Most people would probably agree that to pay or put any pressure on a woman to become an incubator – a thing – would be abhorrent. On the other hand, if she offered to help a childless couple in this way, far from being degraded it could be said that she was making a profound gift to the couple. Yet can she really know when she offers the gift what it will involve for herself? The carrying of the child in her body, and its birth, still lie ahead. Until we know just how these techniques might alter the whole fabric of the parent-child relationship – and above all what it might do to a child if he discovered his extraordinary origins – I for one would not like to see them used. While with luck and sensitive handling adopted children can accept the fact that their true (i.e. genetic) mothers loved them but could not keep them, these children would have to accept that the woman who carried and bore them did so deliberately but without loving them. If this process became accepted, we should have to redefine for our children the whole 'where do babies come from?' question. Above all, we should have to stress genetic parenthood and play down much more important physical and psychological, visible and therefore comprehensible, aspects of pregnancy and birth.

## Rung 11. Gonad grafts

Forget about old men and rejuvenating monkey glands, but remember the two serious technical problems of egg grafts: the chance of damage through handling the egg, and the heavy imposition on the donor. Both of these could possibly be overcome by grafting a complete or partial ovary on to the wife. She would then live with the gonads and genes of another woman inside her, carrying as it were a built-in donor supply. (There is no need to do this with male gonads as long as sperm donors are willing to donate.)

Ethics apart,* this is very unlikely to happen, if for no other reason than that donors would be very hard to come by. Women volunteering for sterilization might be willing to undergo the necessary operation, but they tend to be older and their ovaries less biologically acceptable. Therapeutically removed ovaries are another conceivable source, but then ovaries are not removed without reason.

A more likely possibility, which several biologists are working towards, is the freeze-banking not of eggs but of the egg-producing cells themselves – the oogonia. The only sources for them would be aborted foetuses under six months, since at six months the oogonia have turned into primitive eggs. Assuming it is technically possible to test-tube culture oogonia and harvest ripe eggs from them, there could be an everlasting supply for egg insemination with less risk of damage. However, one technical problem – how to match physical characteristics and so on with a foetus – may well make this clinically unacceptable. Yet it is a striking thought that, if applied, the killing of an unwanted

---

*As for ethics (and the law) there is the curious case of the rich Neopolitan who in the 1930s bought a testis from a Neopolitan boy and had it grafted on to himself. The public outcry was fearsome and led to the drafting of the Italian transplant laws, which are singularly hard on transplanters. I think there would be a similar reaction now, in any country.

female foetus would allow life for another child – its 'daughter' or 'son' – perhaps a thousand miles and a score of years away.

## Sex choice

The power to choose the sex of our children has been hovering about like a ghost ever since we stepped on Rung 2 (A.I.H.). I have ignored it until now because, like all the best ghosts, it makes chilling noises. The prospect of everyone choosing boys or girls at will, and so threatening to unbalance the critical sex ratio, scares many people. Let us now approach the ghost, grasp it, and see if it is substantial.

Briefly, sex is determined at conception by the male sperm. Roughly 50 per cent of all spermatozoa contain a 'female-determining' X sex chromosome while the other half have a slightly smaller 'male' Y sex chromosome. When they meet the X sex chromosome of the female egg the result is either XX (girl) or XY (boy). The simplest sex choice method therefore depends on learning how to separate sperm by any minute 'behavioural' differences between the X- and Y-carriers. No one has conclusively done this yet, but many lines are being followed – especially their slight difference in weight (they might perhaps be spun-separated or allowed to settle out in liquid) and different reactions to a weak electric field. But whatever the successful separation method – and there is bound to be one soon – the sperm must still be *separated from the man* and the chosen fraction inseminated into his wife. So this means A.I.H. – and an emotional and financial barrier that would deter all but the most determined.

Meanwhile the only reliable sex choice method that has been achieved in mammals involves embryos, not sperm. In April 1968 Dr R. L. Gardner and Dr Robert Edwards (see Rung 8) announced in *Nature**\** that for the first time ever they had controlled the sex of the offspring of a mammal. Briefly, what they did was this. They flushed out embryos from female rabbits when they

\**Nature*, 218 (1968), p. 346.

were at the 4,000-odd-cell stage of development. Working under the microscope with micro-tools and astonishingly steady hands, they snipped off about 300 cells from the balloon-shaped skin of cells surrounding the embryos, taking care not to touch the embryos proper. Then they tested these snipped-off samples for their sex. This was possible because, when dyed, all female cells show a small fragment of material called sex chromatin, while male cells do not (though since then it has become possible to detect male Y chromosomes by a simple test with fluorescent dyes). Having established the sex of the samples, and therefore of the embryos themselves, Gardner and Edwards re-implanted the embryos, with their sexes now known, into female rabbits – a male here, a female there – and waited for them to come to term. Their results were remarkable. Though not all the embryos did develop, of those that did the predicted sex was right in every case, and out of twenty all but one was perfectly normal. The exception had no head – probably because the embryo proper was damaged when the test cells were snipped off.

The risk of damage apart, this technique will not be practicable with humans until test-tube fertilization is possible (Rung 8). The reason is very simple. Humans normally produce only one fertilized egg at a time, so that there is no 'pool' from which to choose the sex. With test-tube fertilization, though, several eggs could be flushed out from a wife, fertilized by her husband, cultured in the test-tube until they were mature enough for testing, tested, and an embryo of the desired sex implanted in the wife, there to develop into a boy or girl. Here again, very few people would want to go through or be able to afford this kind of performance – especially the millions in Asia who believe they will go to heaven only if they have a son.

Neither are many people likely to go for the alternative of selecting the sex of a child by aborting any foetus of the wrong sex. This horrific possibility has become possible from research by Dr Landrum Shettles, of the Presbyterian Hospital, New York, which shows that it is an easy matter to determine the sex of a foetus in the womb. In March 1971 Dr Shettles reported the discovery that if one takes swabs of cells from the cervix of a pregnant woman, some of the cells will be from the foetus. It is a simple matter to test a batch of cervical cells with the fluorescent dyes that pick out male Y chromosomes: if any male cells are found, they must come from a male foetus. If no Y chromosomes are found, it is highly likely that the foetus is female. Using this technique on 30 women at various stages of pregnancy, Dr Shettles reported complete success in predicting which foetuses would be boy babies and which girls.

Meanwhile the prospects of a sex-choice pill are extremely remote. Though there are chemical means of influencing sex – for example, the pioneer of contraceptive pills, Professor Gregory Pincus, once managed to produce litters of rabbits which were 90 per cent female by immersing rabbit eggs in oestrogens – a chemical pill that has sufficient potency to influence sex *reliably* and does not produce severe side-effects hardly seems possible at present.

But assuming there is some method of sex choice, what difference would it make? The awful truth is that we hardly know. The crucial surveys, which would reveal which sex couples would choose for their next child at any stage of family-building and whether they really would bother to choose, have never been made. So we are left with guesswork, of which there is no shortage.

Many demographers guess that there would be little difference, or a slight preponderance of boys. They seem to be supported by the few crude surveys that have been done. For instance, the largest one carried out in Britain,* which took a random selection of 20 per cent of all marriages in the city of Hull between April 1965 and March 1966, found that most of the newlyweds (45·7 per cent) wanted to have a family of two with one child of each sex. Two boys and one girl was the next most popular choice (15·4 per cent), followed by four children with two of each sex (12·6 per cent). If all the couples who were interviewed were able to achieve the desired size and sex of their family, they would have had 491 boys and 423 girls – giving a sex ratio of 116 boys

* Peel, John, *Journal of Biosocial Science*, 2 (1970), p. 45.

to 100 girls. The prevailing sex ratio is 105 or 106 boys to 100 girls.

Other demographers guess that many parents would *plan* to choose a boy first, then a girl and leave a third to chance. This would produce an excess of boys because having got their first planned boy most couples would then have their other children naturally – and get roughly half boys, half girls. Or again, the first one or two children might be natural and the *last* selected if necessary. To me this seems far and away the most probable pattern, especially if selection meant A.I.H. and cost a hundred pounds or so. It is, after all, how most parents think about the problem: one only thinks about 'balancing the sexes' when one has some children to put in the scales.

Working from this last model, the table below shows how much different choice levels for selecting boys *only* could affect the sex ratio – assumed here to be 105 boys to 100 girls at birth, about an average figure. The numbers show how many boys would now be born for every 100 girls.

TABLE 3.1   What sex selection might do to the sex ratio

|  | Every couple uses sex selection | 50% of couples do | 20% of couples do | 10% of couples do |
|---|---|---|---|---|
| Every family has just 2 children. Sex selection is used only when the first was a girl to ensure that the second is a boy. | 170 | 132 | 116 | 110 |
| Every family has just 3 children. Sex selection is used only when the first two were girls to ensure that the third is a boy. | 123 | 114 | 108 | 107 |

With larger families selection to ensure one boy has a smaller effect. It is also much more likely to be cancelled out by 'all boy' families wanting a girl.

In practice, only the numbers in the right-hand half are at all likely, and these are of course far too high, since we have assumed no selection for girls at all. But even with this extreme bias for boys, the sex ratio at the age of marriage – which is what matters – is not affected by very much more than it already is by other factors at work in medicine and society.

Since males are more fragile, the 105-males to 100-females sex ratio at birth gradually changes throughout life. In the past the excess of males disappeared after five years and then reversed, hence the glut of Victorian spinsters – a hundred years ago the sex ratio at the age of marriage was about 90 men to 100 women. By 1900 better medical care was keeping more boys alive and the 'equal ratio' was in the late teens. Today it is around thirty. In other words, medicine has turned the tables and produced a slight excess of bachelors (101·2 to 100 in England in 1960 in the twenty to thirty age range).

The point is that the balance has already tipped sharply without anyone but some demographers being very upset about it. This does not mean, though, that the consequences are not important.

The American demographer, Kingsley Davis, has suggested that the most likely consequences from an excess of bachelors are: a rise in the age at which men marry; some increase in the number of men who never marry; and some rise in prostitution and homosexuality.* Others predict that men will raid younger and younger female age groups for mates. Thus the effects of an imbalanced sex ratio would not all be felt by one sector of society, but would be fairly widely distributed – a non-apocalyptic conclusion that seems to have been supported by what happened in the U.S.S.R. and Germany after the Second World War (sex ratio reached 77·7 males to 100 females in the latter) and in Israel in the early male-dominated period of immigration.

These 'costs' may be too high a price to pay for the benefits of sex selection, or they may not: we have to debate that. But if the costs are thought to be very high, there seems to be no reason in principle why the use of sex selection should not be controlled. It would be very easy for a state to see a swing in boy (or girl) production developing. Then, provided everyone realizes *in advance* that, just as we are free to drive but are restricted to one

* See A. Etzioni, *Science*, 161 (1968), pp. 1107–12.

side of the road, this is a freedom that the state must be allowed to restrict, the swing could be checked by forbidding sex selection to certain groups of people. In other words, sex selection would have to be licensed in some way, probably on a sliding-scale system of priorities and needs. At first, for example, licences might be handed out freely to all. Then if a boy or girl boom seems to be developing (this could be known within months) licences would have to be withheld from the least needy cases – how far up the scale depending on the strength of the boom. The priority scale would not be hard to construct. First-child selection, for example, might go at the bottom. At the top there would be the couple where the wife is a carrier for a sex-linked genetic defect like haemophilia, since they have a fifty-fifty chance of bearing affected sons but no risk with daughters, followed by families with 'same-sex runs' of eight, seven, six and so on, who wish to have a child of the other sex.

If such a system is set up and known by everyone *before* sex selection is possible, then when it becomes possible it can only expand our biological freedom. We will be offered a new but wisely limited liberty. But if not, if restrictions are imposed only *after* we have all come to accept choice as a kind of divine right, then there might be fearful political storms. Governments would act reluctantly, slowly, and probably too late to avoid any booms that did occur. There are few biomedical areas where calm, widespread public discussion is more urgently needed than this.

## Rung 12. Clones, or carbon-copy children

At Rung 10 we reached the ultimate limit of A.I.H. and at Rung 11 the last word in A.I.D. Now we leave sexual reproduction altogether and consider the propagation of people – and animal-human hybrids – like a gardener grafting roses.

In June 1966 a team of Oxford biologists announced in *Nature**
that they had grown seven frogs from the intestine cells of tadpoles. In other words, they had achieved with an animal what

*J. B. Gurdon, *et al.*, *Nature*, 210 (1966), pp. 1240–1.

is routine in the garden: the production of a new organism from a cutting of a single parent. In outline, what they did was a gene swop. They took a frog's egg, removed its gene-bearing nucleus, and replaced it with the nucleus from a body cell of another frog (or rather, tadpole). They then implanted the new nucleus plus egg in a female frog and induced the whole system to start the egg developing. The result was a frog. But because the grafted nucleus came from a body cell, which contains the complete *double* set of genetic instructions of its owner, the resulting frog was a genetic replica – a true identical twin – of its single parent. If one does this several times using the same donor, the result is a whole caste or 'clone' of identical twins, who are also identical to their own parent.

Since Gurdon's initial experiments the techniques of nuclear transplantation have been refined considerably, so that success rates are much higher. Other, much simpler, methods of achieving the same ends have also been found. The most important of these is 'cell hybridization'. Using certain viruses it is now a routine matter to persuade cells of different species to fuse together into a single, viable hybrid cell from which cell colonies or clones can be grown. Each cell in the clone will contain genes and chromosomes from each of the parent species used to make the hybrid (see Rung 13). However, if an egg cell is hybridized with a body cell of the same species the genetic material of the egg cell is made inactive. The new hybrid cell is controlled entirely by the genes and chromosomes of the body cell.

Because of these advances, several biologists are sure that if anyone cared to try the experiment it would be possible now to make identical twin clones of a man or woman.

Possible, but how likely? More arrant nonsense has been written about the hopes and dangers of cloning than about any other biological technique. At one end of the scale there is the full-fledged, scare-mongering fantasy about despotic rulers turning out unlimited numbers of carbon-copy people from a Master Clone. For instance, in *The Biological Time Bomb* Gordon Rattray Taylor* has suggested that the free world may be forced

*Thames and Hudson, London, 1968.

into state-controlled cloning to compete with the Yellow Peril: 'If an Oriental despot should decide that he could produce more rugged soldiers, more brilliant scientists, more skilful workmen or more fertile women by such techniques, he might pour the necessary resources into making them practicable, and then impose them.'

*Impose them?* Really? For five thousand years it has been perfectly possible for power-mad rulers to breed selectively and cull human beings into specialized types, just as humans have bred cows and dogs. None have ever done so because none of their subjects would stand for it for a moment and, anyway, if one wants to make specialized men it is far easier and quicker to train them. Enforced cloning would be just as difficult and pointless.

Several biologists with only slightly less fevered imaginations have foreseen a benevolent, voluntary mass use of cloning to improve the human genetic stock. It could be used to 'fix' good characters. This, after all, is why plant breeders take cuttings. They use sexual reproduction for genetic experiments – for mixing genes up to produce chancy, favourable recombinations – and then when a good variety appears they fix it by grafting carbon copies. So why not humans too? As Professor Joshua Lederberg, the American Nobel-laureate biologist has written, 'If a superior individual – and, presumably, genotype – is identified, why not copy it directly, rather than suffer all the risks, including those of sex determination, involved in the disruptions of recombination.'*

In other words, when an Einstein or Russell appears, why not ensure more Einsteins or Russells? One needs only a scraping of cells – from inside the mouth say – to get hold of their 'superior' genes and with freezing they could be used indefinitely, long after their original owner's death. To get over the problem that Einsteins and Russells are not entirely products of their genes, several biologists take the idea further. J. B. S. Haldane pre-

* In his important article, 'Experimental Genetics and Human Evolution', in the *Bulletin of the Atomic Scientists* (October 1966), pp. 4–11. All Lederberg quotations in this section are taken from this article.

dicted* that 'superior individuals' will spend their later years educating their own clones. Because the whole clone school – teacher as well as children – would all be identical twins, they would interact almost telepathically and go very far very fast. A variant on this theme, from Lederberg, is that clone-pairs might be extraordinarily skilled in stressful jobs – as a pair of astronauts, for example, or deep-sea divers, or surgeons. Haldane suggested producing even more specialized clones, for example to preserve rare qualities like great longevity or extra good vision. Other biologists have talked wistfully of the clone football team or bridge pair.

Ruling out coercion, can one soberly imagine any couple freely choosing to conceive a child by these extraordinary means just to be able to hand it over to be educated into genius, or a group of parents deciding together that it would be so nice if they all cloned from a famous sportsman and had their carbon-copy children reared together so that they could make a good football team in twenty-five years' time? People have children, their children, not specialized tools.

Nevertheless, there are some possibly valid and likely reasons for cloning. As Lederberg has suggested, the prevention of mental and physical handicap by eugenics, genetic surgery and biological engineering (see Chapters 7, 8, and 9), though splendid if it can be achieved, may be only a partial answer and anyway is a long way off. Cloning might be nearer and would be surer. For example, if one partner in a marriage has a severe genetic defect the other could be the clone-parent. If both carry the same severe recessive gene defect, so that though outwardly normal they face a one-in-four risk that any natural child will actually have the defect, then each parent could take it in turns to produce a carbon-copy child. Cloning would also be a sure way of selecting sex. In this situation – inside marriage and with the aim of avoiding defects – it is far more likely that a man could persuade a woman to bear his carbon copy. Women, of course, do not face this problem of finding a willing womb: they can simply

* 'Biological Possibilities in the next 10,000 years', Gordon Wolstenholme, ed., *Man and his Future* (J. & A. Churchill, London, 1963).

clone themselves. However, even here cloning is not all that likely. As we have seen on the earlier Rungs, there are less extraordinary ways of not passing on one's own genetic defects.

Still, it is conceivable, and we ought to be aware now of any problems it might raise. Lederberg and other biologists have pointed out several biological dangers. One is the broad, genetic fear that clonishness – the tendency to clone – will become self-perpetuating. By copying themselves, people with genetic defects (or abnormal narcissism or self-esteem) will produce children who will copy themselves and so on. And even though it may produce superior types, vegetative reproduction is a rigid, evolutionary dead-end. There is also the individual, medical risk that we have met before: the problem of the single, damaged egg. But despite this, biologists are far from unanimously against it. Lederberg says, 'My colleagues differ widely in their reaction to the idea that anyone could conscientiously risk the crucial experiment, the first attempt to clone a man. Perhaps this will not be attempted until gestation can be monitored closely to be sure the foetus meets expectations.'

For my part, this is one experiment at which I would draw the line without hesitation, foetal monitoring or not. For biological considerations are not all. Clonal reproduction introduces something totally new into the world – the mind of a child who *knows* it is a biological replica of its parent, a child who knows it is largely preordained, a freak who can see its biological future mirrored in another person.

If this other person was famous it could be psychologically crippling. To aspire to genius is fine, to be the child of genius can be dreadfully difficult, but to be expected to develop into genius because you are its identical twin could be crushing. Lederberg has written of cloning that 'we would at least enjoy being able to observe the experiment of discovering whether a second Einstein would outdo the first one'. The cloned guinea-pig might not share the merriment.

Even if cloning were kept in the family, it could be psychologically disastrous. Children whose fathers insist on seeing and rearing them as 'chips off the old block' usually have extreme difficulty in escaping from their fathers enough to find out what they are really like. A literal chip off the old block would not stand a chance.

## Rung 13. Man-animal hybrids

The clone child can only come with a sudden leap – a first experiment. The 'subhuman' hybrid could creep up on us. Already in several biology laboratories around the world there are cultures of cell tissue which are genetically part man, part animal. There are man-ape cell cultures, and man-mouse cultures. These are semi-formless masses of body (somatic) cells and there is no chance that they will ever stand up and grunt. They are merely extremely useful biological tools for investigating such things as cancer and immunity.

But biologists are restless characters and one day one of them will slip a fragment of human genetic material into an animal egg cell and produce the first man-animal hybrid. As Lederberg has said, 'Before long we are bound to hear of tests of the effect of dosage of the human 21st chromosome on the development of the brain of the mouse or the gorilla.'

Would anyone mind? Not with a mouse perhaps. Not with a gorilla, even, if it was just one chromosome. But what would we think of a true half-and-half man-primate chimera? Would he be a man, or an ape?

Lederberg, echoing the feeling of many biologists, urges us to keep calm. We should start *now* the public debate on what it means to be human, and he suggests, 'Rather than superficial appearance of face and chromosomes, a more rational criterion of human identity might be the potential for communication within the species, which is the foundation on which the unique glory of man is built.'

The advice is sound if the definition is not. But as we stand on this 13th Rung of the ladder that started so simply with a visit to the infertility clinic, I cannot help feeling that we have left calmness and reason far behind. However blandly rational biologists urge us to be about defining the humanity of a quasi-human hybrid they have produced, however neutral or beneficial their intention in producing it, I do not believe most people will see their point.

They will demand a very old and simple criterion for humanity: that a human being is born of a woman. With a hybrid this can hardly be so. Unless one can seriously contemplate any woman voluntarily submitting to the manipulations that would give her a man-animal child, such a hybrid must be born of an animal. Consequently, it is not of our species and does not command our species loyalty: if we had to choose to save it or the most incomprehensible, strange-looking human aborigine from a fire, we would choose the latter. And yet if the hybrid is of another species, can we tolerate the manipulations that blur the gap between animal and man? Biologists may have shown us that we are closer to our non-human origins than was ever supposed even after Darwin dropped his bombshell, but deep traditions insist on drawing a firm dividing line nevertheless. If a man-animal hybrid is produced we will keep it in a zoo or laboratory, not our homes, and pay to see it and embarrass ourselves. But because of what has been done, it might embarrass us more than we can stand.

Biologists working in this area should perhaps consider the myth of the Minotaur, the bull-man that had to be locked in a labyrinth because it was too awful to gaze on.

# 4. BREEDING FOR QUALITY

A century ago Charles Darwin's cousin, Francis Galton, had a vision. When biologists knew enough about genetics, he dreamed, humans could begin to breed as they had long bred animals. By deciding who breeds with whom, who should breed much and who not at all, the human race could improve itself. It could produce new generations where more children have 'good' qualities and fewer have 'bad' ones. And for this ambitious scheme of self-improvement he coined the name 'eugenics'.

It is no secret that biologists *are* now learning so much about the genetic and molecular machinery of life that eugenics and much more of the same kind are fast becoming realizable. We do seem to be on the threshold of astonishing powers of controlling not just how many children we have but what kind we have. We are learning to become *quality* controllers as well as quantity controllers, and the implications – for the evolutionary prospects of the species, for society, and for individuals – are staggering. Therefore, even though some of the prospects are still fairly remote, it is high time that we all talked about them in an informed way. There are obvious dangers in these new powers, and some real opportunities for reducing misery. If we don't all face them now, calmly and rationally, it is much more likely that we will run into the first and take longer finding the second.

In this chapter we shall start with the first part of the eugenic dream – the business of breeding to increase overall 'good' qualities. This will help to clear the ground for the more exacting biology and greater challenges of the second part, *negative* eugenics, which aims at eliminating the specific genetic mistakes that cause defects and disease. Then we can go on beyond eugenics to more radical ways of altering man's basic biological character – to genetic engineering, or biological engineering, in which one leaves genes alone but tries to correct their effects with the precise molecular tools of tomorrow's defect-medicine.

Meanwhile, is positive eugenics possible? Could we, if we wished, give to more of our children the qualities most of us admire: high intelligence, say, or beauty, stamina, compassion, longevity and impressive physique? Briefly, the answer is 'yes' – but it is very unlikely that any community would bother.

For the moment, we shall leave the first objection – how one can agree on what is admirable – and deal with the second – that most admirable qualities are not genetically inherited. Few biologists would go to the stake asserting, for instance, that the Bach family was thick with musical genes or that Gandhi inherited his value system at conception. But there are several important qualities where genes do at least set the upper and lower limits of the potential on which environment and upbringing can work, including 'beauty', stature and, of course, intelligence – or rather what we measure as intelligence in an I.Q. test.

Most of these character qualities appear to be set by many genes working together. They are multi-factorial or polygenic. It also seems that we can hand on different-sized chunks of our genes for them intact at conception, where they are mixed with those of our husband or wife. So as regards these characteristics our children can largely resemble ourselves, or our mate, or be any blend in between. The most likely combination, though, is a roughly fifty-fifty blend, with more and more unbalanced combinations being less and less likely. In other words, an I.Q. 100 wife and an I.Q. 120 husband are marginally more likely to have an I.Q. 110 child than anything else, but it is not much less likely that they will have children of I.Q. 100 or 120, and not very much less likely that they will produce an I.Q. 90 or 130 child. We often throw up 'sports'. However, with a whole population all these individual variations are smoothed out, so that *as long as mating is random* the new generation has the same I.Q. variations as its parents. It is populations, not people, that breed true.

This is crucial. Because geneticists cannot sensibly talk about individuals the whole positive eugenic argument is almost invariably discussed in terms of populations. This means, inevitably, that it sounds authoritarian. It creates a Them and Us atmosphere: 'If *we* wanted to boost I.Q. . . .', 'How can *we* decide what qualities are desirable?' and so on. Individual choice has to be smothered due to the complexity of individual heredity, and you get the dreadful taint of the *pogrom*.

Let us accept this for the moment, but bear in mind that any eugenic programme can be voluntary. What we now have to ask is what happens when the mating pattern or the breeding pattern of a population is made non-random, or directed. We shall find that it all depends on how aggressively it is done; or as Lederberg has put it, 'Positive eugenic programmes can be defended roughly in proportion to their ineffectiveness.'

Keeping to I.Q., one can imagine a very unaggressive state programme whereby bright couples are 'encouraged' to meet and marry, perhaps by providing as many university places for women as there are for men. But this would have no eugenic effect whatsoever if the brights married and bred among the rest of the population. Every I.Q. 140 man who marries an I.Q. 140 woman merely deprives an I.Q. 100 woman or man of the chance of marrying 'up'. The total pool of 'I.Q. genes' is not altered. Marriage by itself can only have a eugenic effect if a group of people decide to isolate themselves from the community.

Suppose a group did just this and started an I.Q. cult with intense selective breeding for intelligence. Would they produce a new race of super-eggheads? Kenneth Mather, the British geneticist, has estimated that with the most intense selective breeding imaginable a group of humans could double (or halve) their average stature or I.Q. in no more than twelve generations.* Another British geneticist, John Maynard Smith, has produced a more imaginable possibility.† He supposes that if most women

---

* *Human Diversity* (Oliver and Boyd, Edinburgh and London, 1964), an extremely readable, comprehensive semi-popular account of population genetics.

†'Eugenics and Utopia', a chapter in the special Spring 1965 issue of *Daedalus*, the Journal of the American Academy of Arts and Sciences.

in an isolated group had at least one child by a high-I.Q. mate, after three generations the mean I.Q. of the group would have risen by 15 points. If they started as average they could get most of their great-grandchildren into university. As Maynard Smith remarks, this hardly seems to justify the establishment of a new cult or religion. What if they went on like this not for three but for thirty generations? Would they raise their average I.Q. by 150 points? Almost certainly not. From animal breeding it seems that the response to intense selection for any character slows down and eventually stops at a certain level. Though no one can predict where this would be with human intelligence it could be high. Our I.Q. cultists could perhaps raise their average I.Q. by several jumps of 15 points and produce many individuals with mental capacities far, far greater than those of anyone alive today. Whether they could achieve anything, or avoid neurotic breakdowns, is a different matter. We hardly know, since we are only just learning how to cope with today's highly gifted children.

But there is a deeper danger lying ahead for our I.Q. cultists. As all animal-breeding experiments show, selective breeding rarely gives something for nothing. You can only breed for one character at the expense of others: you can breed race-horses *or* cart-horses; you can breed pigs for a good bacon yield, but they end up stupid, docile and unable to walk more than a few yards. In other words, breeding for all-round 'improvement' is out. We cannot know whether future men and women will accept this dreadful limitation for the sake of boosting their I.Q. (or any other characteristic), but it does seem unlikely. Most people would surely agree with the British Nobel-laureate, Sir Peter Medawar, when he says, 'The genetical manufacture of super-men by a policy of cross-breeding between two or more parental stocks is unacceptable today, and the idea that it might one day become acceptable is unacceptable also.'*

Really vigorous who-marries-who programmes are therefore

---

*This quotation and the argument of the whole paragraph is from Medawar's 'Science and the Sanctity of Life: An Examination of Current Fallacies', *Encounter* (December 1966), pp. 96–104.

out of the question, and if one does not like the idea of separatist eugenic elite groups all positive eugenics-by-marriage is out too. This leaves us with programmes based on how many children one can have. Should there be tax incentives and so on to encourage the more socially desirable to have more children, and the less less?

Many well-meaning people (and many more malicious ones) believe there should be, and for all kinds of reasons. But probably the single most insistent reason is the widespread fear that I.Q. is declining. I.Q. does to quite a large extent go with class, wealth and status (what else is social mobility about?) and the idea is that the middle and upper social or income brackets are much more 'responsible' about limiting their family size, while the feckless thicks are not. Until a few years ago, article after article appeared deploring this differential fertility by social class with dire warnings about the consequences of the dims outbreeding the brights. But it now seems that for two reasons the warnings were unnecessary.

The first reason is that all the early measurements of how low-I.Q. families were breeding swarms of children were wrong. They looked only at large low-I.Q. families. They left out of account the families of the brothers and sisters of low-I.Q. parents who remained *childless*. When they are reckoned with, the dreaded decline disappears.* The second reason is that in some rich countries higher-I.Q. parents have started to breed more. For instance, in one U.S. study it was found that between 1935 and 1940 college graduate couples were breeding at a rate 48 per cent below that needed to replace themselves, but by 1945–50 they were 10 to 15 per cent above replacement. The latest figures from the U.S. confirm this trend.

Of course the state could push this trend even further, as many would like. There is hardly a tax, bonus scheme, insurance policy or wage demand that does not have some (unintended) eugenic effect by making one section of the community richer and so more able to have more children. One only has to direct

*J. B. Higgins, E. W. and S. C. Reid, 'Intelligence and Family Size: a Paradox Restored', *Eugenics Quarterly*, 9 (1962), pp. 84–90.

this to help the chosen elite – teachers, dons, business executives, professionals, the socially prominent or whoever they may be – and there will be a net eugenic drift. And there of course is the fatal double flaw. Any money-based programme must be socially unjust, because it discriminates against the excluded non-elite; and this apart, how can one know where the drift is heading? To keep or achieve social prominence must' call for some ability, but, as J. B. S. Haldane put it, the bad qualities one needs are quite as striking as the good ones.

In the end, every large-scale *directed* scheme for breeding New Men falls down on these grounds. For it to work, the incentives must be large, which makes them socially offensive, and some infinitely perceptive Eugenicist-King must sit at the steering wheel and be absolutely sure that he knows where he is going. And he must also know that the new men he is creating want to go there too.

However, there is another approach: namely, that each couple assumes the role of Eugenicist-King and chooses to have one or more children by an outstanding sperm or egg donor (despite all the social and psychological problems discussed in Chapter 3).

This bold scheme of positive eugenics was first proposed by the American biologist Hermann Muller in 1939, who called it eugenics by *germinal choice*. Later enthusiasts – notably Herbert Brewer and Sir Julian Huxley – gave it other names: *eutelegenesis* and eugenics by *pre-adoption*.*

Briefly, the idea is to use today's donor-insemination practices as the thin end of a wedge in a deliberate, widespread eugenic

---

* Muller first summed up his scheme in his book *Out of the Night* (1935) but later revised it in a stream of articles. The most important are: 'The Guidance of Human Evolution' in *Perspectives in Biology and Medicine*, 3 (1959), pp. 1–43; an abridged but more accessible version of this called 'Should we weaken or strengthen our Genetic Heritage?' in *Daedalus*, Summer 1961 issue; and 'Means and Aims in Human Genetic Betterment', a chapter in T. M. Sonnenborn, ed., *The Control of Human Heredity and Evolution*, (Macmillan Co., New York, or Collier-Macmillan, London, 1965). Huxley's main statement is contained in 'Eugenics in Evolutionary Perspective' – see for example his *Essays of a Humanist*, (Penguin Books, Harmondsworth, 1966).

programme by voluntary choice. Instead of pretending that A.I.D. is natural by matching husband and donor so that everyone thinks the husband is the real father, seize this golden opportunity and try to create 'an especially worthy human being'.

What Muller proposed is that sperm (and, later, eggs) from a huge variety of donors should be stored in banks. As far as possible donors would be chosen whose lives and achievements have shown they have 'outstanding gifts of mind, merits of disposition and character, or physical fitness'. They would be fully documented in a kind of human seed catalogue, and to avoid the danger of instant fashions – an unseemly scramble for the sperm of film stars, pop groups, athletes, or *Time's* Man of the Year – there might be a ban on use until twenty years after the donor's death. Muller believed that once infertile couples led the way, normal couples would one day come to prefer germinal choice for at least one child and 'the obvious successes achieved by this method would within a generation win it still more adherents. It would constitute a major extension of human freedom in a quite new direction.' Present taboos against the idea, he believed, would soon fade. The main psychological barrier – the violation of every father's wish to survive somehow through the genes of his children – Muller calls a mystique. In future it will be superseded by the pride a 'love father' of a germinal-choice child would feel from knowing he had given 'his' child the best possible genetic endowment. We will all become more rational about inheritance. We will condemn as a childish conceit our present fixation on passing on our own genes with all their 'peculiarities, idiosyncrasies and foibles'. Instead we will find fulfilment in passing on the *best* that we can find to represent ourselves; and this will represent a higher form of morality.

Perhaps our grandchildren may see it this way. But at present the majority of geneticists doubt it because germinal choice must always be hopelessly ineffective – for individual couples and for a population. There are simply too many biological arguments against it, let alone all the obvious personal and psychological ones.

Take individual couples first. Most worthy donors would be chosen when they are middle-aged or older and the seed catalogue would not list their accumulated genetic damage. No catalogue could sift environmental from genetic factors in the donors, and even if it could, the uncertain nature of reproduction would very, very often dilute out the 'good' genes. Very many couples, in short, would be sorely disappointed.

The most damning argument, though, is that germinal choice would have a negligible eugenic effect unless it became extremely fashionable. Maynard Smith has estimated* that if 1 per cent of average-I.Q. women had half their children by donors with an average I.Q. of 115, in a generation the average I.Q. of the population would rise by 0·04 points. Even if 10 per cent opted for germinal choice with donors of I.Q. 160 the rise would still be a mere 1·5 points. It hardly seems worth while, especially since it is now known that sensitive teaching can raise a deprived child's I.Q. score by 15 to 20 points. Better schools do seem to have claims to a higher priority than building sperm banks for better genes.

But there is one last argument from the positive eugenics enthusiasts. Muller and Huxley have insisted that even tiny average I.Q. rises are worth while because they would produce more geniuses at the top. In other words, if the middle of the I.Q. curve is moved a fraction towards the right (bright) side, genetic variation will see to it that there is a big jump in the number of very high scorers. One estimate is that a 1 per cent average rise would raise the proportion with an I.Q. of 175 or more from about 3·5 to 4 per million – a shift of 15 per cent. If one could produce 15 per cent more Einsteins or John Stuart Mills, *that* might be worth while. Other biologists dispute this on the grounds that an extremely high I.Q. is almost certainly not entirely due to the action of many genes. The classic bell-shaped curve of I.Q. distribution is stretched out a bit at each end. Specific 'bad' genes and birth damage produce more imbeciles than one would expect from it, while there are probably specific 'high-I.Q.' genes for producing genius. On this model, the

* See footnote p. 119.

number of people with an I.Q. over 175 is not about 3 per million but nearer 80, and a 1 per cent average shift would increase this to only 85 – a tiny proportional increase. Besides, as Maynard Smith has argued, why suppose that more geniuses would give more great works of genius? Why did Periclean Athens produce so much when Byzantine Greece did not? You may need unusual genes to be an Einstein, but doubling their number will not necessarily produce twice as many relativity theories. Genius will only come out in a favourable social climate, and *that* is what we should be aiming at.

No, when we wake up to the cold light of biological and social reality the first part of Galton's dream remains a dream. And let us not get too anxious that it should. There is a more devastating potential in our gonads to think about.

# 5. BREEDING OUT FAULTS

Most of us produce babies with as much idea of what kind we are going to get as a child dipping for a present in a bran tub – and with the same mixture of hope and fear of disappointment. We set a couple of hundred million sperm after an egg whose genes we know just as little about and then wait nine months, or years, to know if we have won or lost the inheritance lottery.

Most of us win. Though we all carry lethal or crippling genes in our gonads, we have normal children. Luck is on our side. But sometimes the genetic fruit-machine presents a combination that spells catastrophe. At least one in every eight established embryos is killed by its genetic defects. One in every forty or so babies is stillborn or dies within a year because of birth hazards or congenital defects. Worst of all, about one in twenty-five babies is born with a physical or mental handicap and *lives*. In Britain alone these living survivors of cruel accidents of inheritance, development and birth arrive at the rate of around five hundred a week and add up to a total handicapped population of about two million. Though by no means all are gravely disabled, together they comprise one of the principal causes of human suffering.

In the next few years biology and medicine are going to give us the chance of reducing this misery. Not eliminating it, but cutting it dramatically. As they lay bare the genetic and chemical machinery of man, they will accumulate techniques that could be used to cure or prevent a great many kinds of birth defect. But whether we seize the chance they offer is a different matter, because before we can we shall have to change quite drastically our attitudes to many traditional practices and 'rights'. Few of these techniques will come without changes, sacrifices and expense. It is not therefore just a matter of novel means; once again bio-medicine is challenging us to think very carefully indeed about ends, and to decide what it is we really want to do.

One of the biggest challenges (and the main theme of this chapter) is that in the near future we shall – or could if we wanted to – know a great deal more about the 'bad' genes we all carry. This is the challenge of negative eugenics. When we have this genetic foreknowledge and *know* the risk we run of producing a defective child, what are we going to do? Ultimately, if we ever achieve the necessary technical wizardry, we may be able to take our genes along to the genetic surgeon and have them 'cured'. But in the meantime the responsibility of this new knowledge will hang heavy, because as long as we have to reproduce with the genes we have, unaltered, genetic foreknowledge will force us to ask ourselves whom we dare marry and have children by, or whether we dare have children at all. And bearing in mind the fantastic cost of coping with birth defects, it may force society to answer the questions for us by 'advising' us what to do.

Of all the hundreds of different kinds of serious birth defect or disease, only one quarter or so are definitely known to be due to actual genetic mistakes. As we shall see in the next chapter, the others are due to something going wrong in the womb or at birth or – and these are the majority – to 'complex' (in other words, unknown) causes. However, as epidemiologists reveal more of these unknown causes, it is fair to assume that a goodish proportion of them will turn out to be purely genetic in origin. The American geneticist J. F. Crow has estimated, for example, that though there are 250 or so gene defects of the type known as recessive there are perhaps another 600 still to be discovered.

In the meantime, that quarter is quite enough to take seriously. It means that about 1 per cent of all surviving babies are the victims of the transmission of specific 'bad' genes, and when a country produces a million babies a year (as Britain does) that means 10,000 new gene-defect victims a year and a surviving population of around 500,000. Further, the proportion is in-creasing relatively as medicine cuts down other killers and illnesses but leaves genetic defects more or less untouched, and also as it prolongs the lives of gene victims who even ten years ago would almost certainly have died. To give just one example, a recent British survey found that genetic or 'partly genetic' reasons for blindness rose from 37 per cent of all cases in 1922 to 68 per cent in 1950.

So that, briefly, is the scale of the problem. What can be done about it depends on what kind of genetic defect is involved; in other words, whether it is a dominant, recessive, sex-linked or chromosome defect. These are the four pillars of negative eugenics.

## Dominant defects

The human genetic blueprint contains several hundred thousand genes, or rather *pairs* of genes. Each gene pair is probably responsible for one of the hundreds of thousands of chemical reactions or chemical production processes that make the body work. Some are 'know-how' genes: they carry the information needed for a cell to string chemical sub-units together to make the myriads of proteins and enzymes that we need. Others are 'know-when' genes, which tell the 'know-how' genes when to start and stop manufacturing. Physically, a gene is a stretch of a long but thin thread-like molecule called DNA that stores information, much as writing does, by means of a sequence of four different chemical groups – the four letters of the genetic language. How it does this is another (and well-told) story. For the moment all we need to know is that genes come in pairs and that some individual genes can be so defective that either they make a badly defective protein which impedes the chemical works or they make none at all.

With dominant defects, it is the first of the two possibilities. One of the genes makes a protein abnormal enough to do serious and often widespread damage, so much so that its effects swamp those of the other gene of the pair. If we call the gene pair $Aa$ we can say that it is the $A$ that causes the defect and the $a$ does not matter. It follows, obviously, that one has the defect only if one inherits an *A*; and less obviously that only people who *know they have the defect* can pass it on (unless it is so trivial that it does not matter anyway).

When we produce sperm or eggs the double gene set is split in two. One gene of each pair goes into one sperm (or egg), and the other into another. And so on for all sperms or eggs we produce. In other words, if we have an *A* gene it will only get into half of our germ cells – and on average into half of our children. This does not mean that if a first child gets the defect the second will not. It means that every child has a fifty-fifty chance of inheriting the defect. The inheritance pattern is shown in Figure 5.1.

FIGURE 5.1   Dominant inheritance

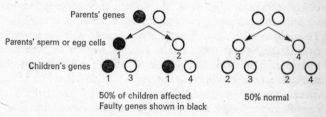

Parents' genes

Parents' sperm or egg cells

Children's genes

50% of children affected
Faulty genes shown in black

50% normal

Many dominant defects involve several genes, and the action of the environment throughout development may also affect how strongly the gene defect shows up as an actual bodily defect. So the inheritance pattern is not always as clear-cut as this.

But, broadly speaking, with dominants there is no present knowledge-barrier to deciding whether or not to have children. The person with a dominant defect knows he has it; he knows what suffering – or otherwise – it causes him; and unless he is mentally subnormal he can understand the clear risks for his children.

Typical examples are achondroplasia (the circus dwarf with short limbs); epiloia (severe mental deficiency often with epilepsy); and retinal aplasia, which accounts for a tenth of all blind babies. There are probably several hundred dominant defects in man, though most of them are still to be discovered and not all are or will be as severe as these three examples. Individually, dominant defects are rare: a frequency greater than 1 in 5,000 births is unusual.

## Recessive defects

There are probably ten times as many recessive defects as dominant ones, and they are mostly much more severe. Many are lethal. More than 250 have been identified and more are discovered literally month by month. Fortunately, most are individually rare – a typical frequency is around 1 in every 10,000 births – but several nasty ones are much commoner than this.

With a recessive defect *both* genes of the pair are defective, but defective in such a way that the protein they control does not actually do harm but simply does not work. In effect, it is missing. The consequences, even when only one gene pair (i.e. protein or enzyme) out of thousands is missing, can be disastrous or dramatic. The baby lacking the enzyme for converting the phenylalinine in his diet into tyrosine will slide down into severe mental retardation. He is a phenylketonuric. The baby who cannot convert tyrosine into thyroxine because he lacks one enzyme will be a goitrous cretin. The baby that cannot do the two-step conversion of tyrosine into melanin will have white hair and pink eyes. He will be an albino. And so it goes for enzyme after missing enzyme, tiny cog after tiny cog of the vastly complex chemical machinery of man.

To have both genes of a pair 'missing' the victim of a recessive defect must have parents who *both* have a 'missing' gene themselves. That is, while the victim is *aa* the parents are *Aa*. They have one normal and one missing gene. But almost certainly they are unaware of it. It seems to be a very general rule that the body can get by perfectly well with only one normal gene or – the same thing in effect – only a single dose of the protein or enzyme that gene controls. But the parents are *carriers* of the recessive trait and when two carriers have children the results are all too predictable, as shown in Figure 5.2.

Though victims of recessive diseases are individually rare, this does not mean that carriers are rare. There is almost certainly not one person living who does not carry several severe or lethal recessive genes: estimates vary from one or two to half a dozen. What usually saves us from disaster is that because there are so many recessive genes, it is unlikely that we will marry someone with the *same* one. Take the albino as an example. About 1 in 20,000 babies is an albino, but the chance that you carry the albino recessive gene is about 1 in 70. The same chance applies to your husband or wife. So the probability that you *both* carry it is not 1 in 70 *plus* 1 in 70 but 1 in 70 *times* 1 in 70: or 1 in

FIGURE 5.2 Recessive inheritance

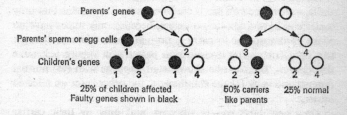

Parents' genes

Parents' sperm or egg cells

Children's genes

25% of children affected
Faulty genes shown in black

50% carriers
like parents

25% normal

4,900. If you are that unlucky, then only one quarter of your children on average will get both genes. So the chance that any couple will have an albino child is a quarter of 1 in 4,900, or 1 in 20,000.

As long as carriers do not know they are carriers they marry and bear children in happy ignorance of the slight peril they are risking. Many couples indeed never know what they have risked. Even though both *are* carriers of the same recessive gene they may have six children without the genetic dice rolling against them and producing an affected child. In this way recessives can be handed on unseen for generations, quite unsuspected, only to appear perhaps a century later as an apparently random accident of birth. Only when that accident has happened – whether to one of their own children or to near relatives – can a couple know that they need genetic advice. But at present the price for this advice – the birth of an affected baby – is high. As we shall see, this is going to change dramatically.

## Sex-linked defects

Common examples of sex-linked defects are haemophilia ('bleeding disease'), red-green colour blindness and a large proportion of muscular dystrophy. They are so called because a recessive gene is carried inside one of the two sex chromosomes of a parent. In women these sex chromosomes are identical and shaped roughly like an X, and women are therefore XX. In men the sex chromosomes are different. One is smaller and Y-shaped, so that males are XY.

All but one sex-linked defect – it causes hairy ear tufts in parts of India – are carried on the X chromosomes. This means that only males can have a sex-linked defect, because only males have no spare chromosomes to carry a normal gene. Women, though, can be carriers. The transmission patterns in Figure 5.3 make this clearer. The top one shows what happens with two normal parents and is by far the commonest; the other is for an affected father.

Thus only boys can be affected, and only by their carrier mother. Conversely, only a normal couple can produce an affected boy, while the affected man cannot produce affected children (though all his daughters will be carriers). As long as carriers cannot know that they are carriers, the only means of prevention is to advise the sisters of an affected boy that there is a fifty-fifty chance they are carriers and therefore have a one-in-eight chance of having an affected son. An affected male can be warned, not for the sake of his children – none will be affected – but for the male children of any daughters he may have.

## Chromosome defects

At least 1 per cent of all liveborn babies have a gross misprint in their genetic instructions. Instead of a single gene (or a few genes) being 'wrong' a whole gene package – a chromosome or part of a chromosome – is passed on wrongly. The effects can be disastrous. Chromosome defects are thought to cause at least a quarter of all spontaneous abortions, and for the survivors the effects are sterility, mental retardation or some form of malformation. Mongolism is their archetype. Yet, oddly enough, the effects of some of these gross misprints are surprisingly mild.

About half of all these defects are in the sex chromosomes.

FIGURE 5.3   Sex-linked inheritance

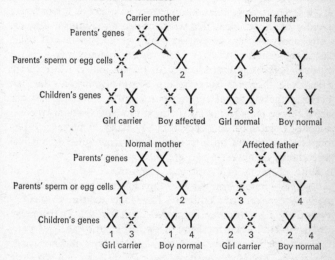

About 2 in every 1,000 male babies, for example, have XXY instead of XY sex chromosomes. These Klinefelter males are nearly always sterile and some also have small testes and look like eunuchs. About 1 in 3,000 girls has only one X sex chromosome (Turner's syndrome) and may suffer from some drastic defects, including grossly immature genital organs, mental deficiency, dwarfing, a webbed neck and deafness. But girls (about 1 in 1,000) who have three X sex chromosomes get off lightly, with a slight risk of mental deficiency and, in some cases, underdeveloped secondary sexual characteristics. In the last decade innumerable variants of these three main types have been found. There are 'female' males with XXXY and XXXXY (gross skeletal defects and mental deficiency) and a whole range of 'double males', XYY, XXYY, and so on.

In the mid-1960s these YY males attracted a lot of attention after surveys in special security institutions suggested that they had a striking tendency to be tall, aggressive, anti-social and delinquency-prone. One study found that of 1,000 males in an institution, 21 were YY – ten times as many as one would expect from the normal population and twice as many as for ordinary mental defectives. News that scientists had discovered a 'gene for delinquency' began to hit the headlines and in several court cases possession of the YY chromosome pattern was accepted as evidence of diminished criminal responsibility. More recent and more thorough surveys, however, suggest that the link between the abnormal chromosomes and abnormal behaviour are far more tenuous than was at first thought. Certainly there is a general statistical tendency for YY males to be tall, feeble-minded and to end up in special security institutions; but just as certainly many YY males are perfectly normal.

With misprints in the other chromosomes – the 22 pairs of *autosomes* – mongolism is by far the most common and serious. About 96 per cent of all mongols have inherited three instead of two of the chromosomes numbered 21. They are '21-trisomics' or 'triple-21s'. Their mothers, instead of splitting their 21 pairs one into each egg, have dealt two into one egg and none into the other. The fathers fertilize the double 21 egg and so add a third 21 chromosome.

Unfortunately, though this looks like a simple inheritance pattern, actually there is no pattern at all. The mother of a mongol seems perfectly normal because the copying error that produces mongolism happens only in her gonads or in the first cell divisions of the embryo. Mongolism and virtually all chromosome defects arise from *hidden*, random mistakes, rather like new mutations.

Because of this, about the only possible preventive measure is to advise older women not to have children. About 1 in 600 liveborn babies are mongols, but while *any* woman under thirty has only an insignificant chance of having a mongol child (around 1 in 2,000) the chance goes up to 1 in 300 for thirty-five- to thirty-nine-year-olds and 1 in 35 for the oldest mothers. This is still a low risk. It is about the same as the risk that any child will be born with a serious abnormality or defect. But it is significant: if women over forty did not have babies mongolism would be cut dramatically.

Apart from abortion or infanticide, the other possible preventive is to cut down virus diseases. There is some striking though still fairly controversial evidence that mongol births come in peaks nine months or so after epidemics of hepatitis, and there is some laboratory evidence showing that serum from hepatitis patients produces mongolism-like chromosome abnormalities in laboratory cultures of cells. Abnormal sex-chromosome babies are also more frequent in poorer families and in spring or summer births, while up to 3 per cent of babies affected by German measles (rubella) have chromosome abnormalities. Environmental factors may play a big part in chromosome defects, as they are suspected of doing in many abnormalities of unknown cause. Only more research will tell. But in the meantime – except for those elderly mothers and the few with defects they know about but who are not sterile anyway, or are not socially sterilized by being locked up in institutions – the New Eugenics cannot do much about 'wrong' chromosomes.

What, then, can it do about 'wrong' genes? At the moment, the answer is very little. Given our *present* genetic knowledge very few couples know they need genetic advice, and so very few have to think about not having children for genetic reasons. But this may not last much longer: soon everyone may have to think and act eugenically.

Couples who need genetic advice need it badly. The vast majority of them have had an abnormal child and want to know the risks of having another. About 90 per cent of couples who go to genetic counselling clinics are in this category. The remainder are mostly middle class, or more educated couples who seek advice before they have had children, or even before marriage. They seek advice either because of some defect in themselves or their families, or because they are cousins.

Once he has diagnosed what kind of defect is present, the genetic counsellor must tell the couple what risks they run with further children. In about 10 per cent of cases he will say the defect is not genetic at all but was due to a pregnancy or birth accident. In that case the risks for another child are the same as for all births: about 1 in 30 to 1 in 40 for a severe or moderately severe defect. In another 30 to 40 per cent of cases he will diagnose a defect of unknown cause. Here the risks of a repeat are also nearly always low – about 1 in 20 or better. Most of his other cases, though, will run much higher risks. There will be the simple dominant defects (risk of 1 in 2 for each child), the recessives and sex-linked (1 in 4) and a few other cases where risks are 1 in 10 or worse. So, broadly speaking, there are two strikingly different risk groups: high risks of worse than 1 in 10 and low risks of 1 in 20 or better.

Equally striking, these two groups roughly correspond to the risks most couples *think* are too high for them to have children or are low enough for them to gamble on. Few studies have been done, but one by Dr J. A. Fraser Roberts is widely quoted as being fairly typical.*

Given a low risk, about 50 per cent of couples are reassured completely about having further children, 30 per cent are left with a niggling worry (though half of them do have more children), and 20 per cent decide not to have more children. This last group, Fraser Roberts suggests, may be using the genetic prognosis as a socially acceptable reason for their not having more children anyway. Given a high risk, about 65 per cent decide not to have more, 15 per cent are doubtful, and 20 per cent decide to have more.

What risks parents will take obviously depends on how severe the defect is. If it is so severe that it will kill the child at birth or after a few months the vast majority of parents will take the risk.

* *British Medical Journal*, 1 (1962), p. 587.

They would rather face that than a crippled survivor.* After that it depends obviously on how life-limiting the defect is and what treatments are available – or rather it depends on how carefully and truthfully the counsellor or family doctor *tells* the parents of these things. Because most counsellors are grossly overworked and few family doctors ever meet any one particular kind of defect, it is sometimes all too easy for them to say, 'Don't worry, we can do wonders with this nowadays,' and forget the long, agonizing effort that the wonders can mean for the child or its family. Many family doctors also assume the role of amateur counsellor themselves and describe recurrence risks as infinitesimal (when they are not) or hand out dire warnings when the risks are low. But given the rapid expansion of genetic counselling in most advanced countries, this situation is likely to change for the better.

The basic challenge of genetic counselling is, of course, whether the parents' decision should be theirs alone, or whether the state should have a part in it. At present virtually all medical opinion says that it must be the parents' responsibility: the geneticist's job is to present the facts and then back out. No one, they say, has a higher right to decide whether it is proper to risk producing a child with such and such a defect than a person who has it already or already has a child with it. For the community to intervene with *its* interests in mind would be an intolerable assault on personal liberty. It would mean compulsory sterilization of the unfit, and that is nasty. Some societies have it. Denmark, for instance, has a eugenic sterilization law (mostly concerned with feeble-mindedness) which leads to the sterilization of about half of all women known to have I.Q.s less than 75: and though many are volunteers, many others are not. But this could not be expanded into a comprehensive set of sterilization laws. How could one possibly draw the line between this defect and that, especially when medicine is all the time improving treatments and even achieving cures?

*Though this may not surprise any parent, it has not got through to the compilers of medical statistics – which almost always measure the gravity of congenital defects by mortality rather than morbidity.

These are strong arguments, but there are powerful arguments opposing them. To my mind the central one comes from what Professor C. H. Waddington has called stochastic (or statistical) ethics. If I deliberately cripple a child I am a monster and the community will lock me up. If a government lined up a thousand babies against a wall and shot them, a political bloodbath would follow. But what if I knowingly take a 1 in 4 risk of conceiving a crippled child? Or kill a child while driving when drunk? Am I a victim of bad luck, to be sympathized with – or a gambler whose rashness is too cruel to be tolerated? And what if a government tests a nuclear weapon in the atmosphere which will, inevitably, somewhere at some time, kill or maim a thousand babies through mutations? Most governments have at last banned atmospheric tests, because the diminishing military gain from them could no longer be said to outweigh public opposition to them, based on their genetic consequences. Several governments have introduced breath tests for drunken driving for similar reasons. No government has yet banned driving, because the risks of injury are too slight when weighed against the benefits. In other words, a large proportion of all social legislation is based on assessing the good of a thing against the risk of its injuring some-one, and then banning it if the injury is greater.

I do not pretend to know how any society (rather than any individual) could balance the 'good' of parenthood – where motives may vary from selfish indulgence to a deep desire for children and the idea that devotedly raising, say, a mongol brings out unplumbed spiritual qualities – how it could balance this against different risks or severities of defect. My point is that the idea that the state should enter into this kind of weighing-up process for the social good is widely accepted in other areas of life. One should also remember that from the technical point of view nearly all the couples who would be banned would soon have sperm and egg donor insemination as an alternative but safe way of reproducing; though how long before there are enough doctors able and willing to apply these techniques is another matter.

If the prospect of state sterilization laws still seems abhorrent, or if one thinks that by giving us greater genetic foreknowledge biological progress will make them more likely, take comfort. For the crucial fact is that when one takes the big step from individual genetic counselling to *population* genetic counselling, compulsory programmes become either largely pointless or an almost trivial intrusion on our 'rights'.

To see why, imagine a society that did have an aggressive eugenic policy aimed at stopping every high-risk couple from reproducing. To have adopted this policy against all objections, the government would have had to be sure that the benefit was worth while. In other words, any negative eugenics policy would have to be effective, whether or not it is unobjectionable. Would it be?

Take dominant defects first. These seem to be the most preventable. No new knowledge is needed: everybody with one knows of it, and knows there is a very high risk of it being passed on. Why not simply sterilize them all? Because, curiously, doing this would have very little total effect. To see why, merely consider the fact that many dominants are lethal. In that case, since their bearer is dead – probably at or soon after birth – how can he pass the defect on? The answer is that he does not: all lethal dominant defects arise afresh in each generation by *mutations*.

Mutation means change, and, with genes, almost invariably a change for the worse. On the basis of the universal rule that if one strikes any complex, finely tuned system with a random blow one is unlikely to improve it, the gene (or DNA) that is struck by radiation or meets a mutagenic chemical or, as happens most often, fails to copy itself correctly, will end up 'wrong'. Most mutations occur in sperm or egg cells and nearly always affect a single gene only. They are also surprisingly common. Estimates of how often they happen vary, but a reasonable supposition is that for most genes from one to ten in every million suffers a mutation. In other words, in the 200 million sperm of a male's ejaculation between 200 and 2,000 will carry a mutation of any given gene. But since only one sperm starts a child the chance of a foetus with a given defect roughly equals the mutation rate.

Because mutations cannot be prevented and can only be reduced slightly by controlling diagnostic X-rays and so on, there is no hope of preventing lethal dominant defects. Much the same applies to severe dominant defects. Consider, for example, the achondroplastic dwarf. Four out of five die in their first year. So if one starts with 100 dwarf babies only 20 will survive to parenthood. If these have 2 children each, half of them – or 20 – will be dwarfs. But meanwhile fresh mutations will occur at just the rate needed to keep the dwarf population the same. That is, 80 normal parents will also produce dwarf babies, but this time out of the blue. So sterilization could prevent the *known risk* 20 – and that one may or may not consider important enough for legislation. But even the most aggressive measures could not reduce the frequency of achondroplasia below the mutation rate, which for this small population means 80 dwarfs (rather than 100) in each generation. At the other extreme, the only dominant defects that *could* be reduced to zero are those where the mutation rate is zero, which means that the defect has no effect at all on fertility or the chance of reaching parenthood. If they are that trivial, aggressive measures are very aggressive indeed.

Unfortunately, there are a few dominant defects which do not appear until middle life. By then the victim has probably had children and so – all unsuspecting – may have passed the malignant gene on to them. Most of these late-developing dominants are mild, but there is one that is very nasty indeed. This is Huntington's chorea, a progressive degeneration of the nervous system that is very distressing and inevitably fatal. When it strikes, the victim's children and siblings – half of whom, statistically, will also carry the gene – must live with the constant fear that they will develop the disease too, or will pass it on to their children.

What could our aggressive eugenic policy do about this? Preventing the birth of choreics *would* drastically reduce the disease because the fact that it develops late in life ensures that the gene is mostly passed on by affected parents and not by fresh mutations. It could sterilize all choreics as they become known, to prevent them having further children, but this would miss the majority who would have been born already. It could sterilize close relatives of a known choreic on the grounds that they have a 1 in 2 chance of getting the disease and therefore a 1 in 4 chance of begetting choreic children. This is a high risk. But there would be more of a case for compulsory sterilization if there was some way of detecting early in life who *will* become a choreic – that is, *who carries the hidden gene.*

Much the same goes for sex-linked diseases. Here at any time one-third of the malignant genes are in affected males and two-thirds are carried hidden in women. There is little point in sterilizing the males because they and their wives cannot produce affected sons. They can only have carrier daughters. And anyway, with severe defects like haemophilia or inherited (Duchenne) muscular dystrophy most boy victims do not live to reproduce. It is the hidden genes in women that do the damage; so again we are back to detecting carriers. If this could be done, and all carriers and male survivors were sterilized, with a severe defect the incidence of the disease would be slashed in a generation to the fresh mutation rate, or about one-third of the old level.

So far we have put our eugenically-minded society in a fairly hopeless position. The more severe the defect the less it can achieve, and the only way it can achieve something is to detect carriers. In fact, that is just what biomedicine *is* learning to do. Carrier detection is one of the hottest new fields in genetics and its implications are far-reaching.

Consider the third type of defect, the recessives. Affected children appear only when two apparently normal carriers of the same gene mate and have children. The 'apparently' is important, because of course they are not normal. They have only one working gene, which means that they make roughly only half as much of the protein that gene controls. The sophisticated biochemistry needed to detect this is now being developed. By mid-1966 there were about twenty recessive diseases where it was possible to detect carriers with a fair to high degree of certainty, and since then there has been a steady trickle of additions, some of them to do with the most common or severe recessive diseases of all.

Here are some of the most important recessive defects where carriers can be detected. The first three are referred to again later.

*Phenylketonuria (P.K.U.).* (Severe mental retardation. Good but difficult 'cure'.) Lacking one enzyme, the phenylketonuric cannot convert phenylalinine, found in most protein foods, into tyrosine. After a few weeks from birth, phenylalinine and other chemicals accumulate, 'poisoning' the brain. On a strict artificial diet containing no phenylalinine but added tyrosine development is normal. Keeping a child on this diet is far, far easier said than done.

*Cystic fibrosis.* (A major child-killer. Treatment is possible but is arduous, expensive and far from successful.) Cystic fibrosis is the commonest serious recessive defect. In white races it occurs in from 1 in 600 births (New York) to 1 in 7,000 (Sweden) with 1 in 1,000 as average. This means that from 2 to 8 per cent of adults are carriers. The symptoms are that the lungs, trachea and intestines fill with a thick, sticky fluid so that the victim literally drowns to death. If he can be diagnosed early the lungs can be kept clear by washing with detergents or, in milder cases, by putting the patient in a mist tent.

With low-fat diets, antibiotics and drugs most children now live past adolescence, but the effort to keep them alive is enormous. In 1968 it was found that victims and carriers have easily recognized dark bodies in cells cultured from their skin. This gives a near-certain, if laborious, carrier test. Soon blood-sample tests might be possible.

*Sickle-cell and Cooley's anaemia.* (Both nearly always kill in childhood. No treatment.) In sickle-cell anaemia there is a minute 'construction error' in the protein, haemoglobin, which colours blood red and carries oxygen round the body. As a result, when the blood-oxygen decreases for any reason the red cells collapse to a sickle shape. In affected individuals nearly all the red cells do this; in carriers roughly a third. By seeing if the cells do sickle, carriers can be detected with 100 per cent certainty.

This is the classic example of a recessive defect that *gives carriers an advantage*. Though getting a double dose of the gene is lethal, a single dose confers resistance to tertian malaria. Since carriers are far more common than double-dose affecteds, the majority is better off at the price of a few sacrifices. In some parts of Africa where malaria is endemic, up to 40 per cent of the population are carriers or affected. In these 'balanced polymorphisms' it is not mutation that keeps the gene in circulation but the clear advantage for any carriers. Once medical progress removes that advantage – for example, by eradicating malaria – the gene slowly declines, as it is doing among African Negroes in North America.

Cooley's anaemia is somewhat similar. There must be an advantage for carriers, but no one is sure what (malaria resistance has been suggested). In parts of Italy it is so common that it has led to the first mass-screening programme for carriers. In Ferrara, for example, a tenth of the population are carriers, and in some villages a fifth are. At this rate, without screening, 1 per cent of all babies would be affected. Instead, in the worst areas, there are intensive screening programmes, with registers of carriers to reduce carrier marriage.

*Duchenne muscular dystrophy.* (Most common and severe type of muscular dystrophy, or degenerative disease of muscle. Most victims are invalids by adolescence, dead by adulthood. Sex-linked. No treatment.) The missing enzyme that causes Duchenne muscular dystrophy has not been discovered, but carriers can be found by measuring the amount of creatine kinase in the blood. This test is far from 100 per cent certain. A high level indicates a carrier, but about one in three carriers seem to have normal levels so cannot be detected. Because of this one cannot prove that any person is not a carrier.

*Haemophilia.* (Persistent bleeding from wounds or surgery, and internally into joints and muscles producing deformed or wasted limbs. Sex-linked. Partial treatment.) The haemophiliac lacks anti-haemophilic globulin (A.H.G., or Factor VIII), one of several chemicals needed for making the blood-clotting agent thrombo-plastin. They can be partially treated by transfusing fresh blood or plasma, or concentrates of Factor VIII. It is virtually impossible to treat often enough to maintain sufficient in the blood, though with great care, special schooling, etc., long-term survival is possible.

*Glycogen storage disease.* (Massive enlargement of the liver; death in infancy or early childhood. No successful treatment.) Normally the body turns glucose sugar into glycogen as a reserve energy supply. With G.S.D. the glycogen is manufactured but cannot be broken down again, so it accumulates in the liver. With frequent small feeds and antibiotics some victims live longer.

*Goitrous cretinism.* (Severe mental deficiency. Total cure in some cases.) Lacking an enzyme, the victim cannot synthesize the hormone thyroxine in the thyroid gland. The thyroid gland enlarges as a neck groitre in a futile effort to compensate. By administering thyroxine throughout life the ill-effects can be avoided.

*Histidinemia.* (Defective or retarded speech; more than half are mentally retarded. No treatment.) The sufferer from histidinemia lacks the enzyme needed for breaking down the protein constituent, histidine. The carrier test is 100 per cent certain.

*Galactosaemia.* (Enlarged liver, cataracts, diarrhoea, jaundice, blindness, mental retardation, probable early death. Total 'cure'.) The galactosaemic baby cannot break down the milk sugar galactose, and so a galactose-phosphate compound accumulates and poisons the body. By withholding all milk the disease can be avoided. The carrier test can pick out about 95 per cent of carriers.

*Maple syrup urine disease.* (Irreversible brain damage. Rare to live beyond one year. Partial treatment.) The patient cannot break down three constituents (leucine, isoleucine, valine) of nearly all protein foods. These accumulate, poison the body, and give the urine a slight odour like maple syrup. A special diet with the three chemicals removed is claimed to have some success in preventing brain damage.

The major technique in detecting carriers is to take samples from the organs where the abnormal chemicals act and see if there is either a deficiency in the *quantity* of protein made, or any

FIGURE 5.4 Detecting carriers of galactosaemia

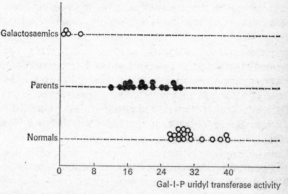

Redrawn from McCusick, Human Genetics (Prentice-Hall, New Jersey, 1964), p. 70.

important abnormality in *structure*. There are now many sophisticated tools for rapidly separating out minute samples of biological molecules into their individual components. They work on the principle that different chemicals will migrate through a gas, powder or even wet blotting paper at different rates. If one passes a mixture of chemicals through them, after a time the mixture will therefore separate into bands, like the slow and fast runners in a race. One can then stop the machine and estimate how much and what kind of chemical there is in each band. Even relatively cheap analysers can separate two proteins that differ by only one chemical sub-group out of hundreds.

For most recessives it is the quantity measurements that matter. Figure 5.4. shows the kind of picture one usually gets – a fairly clear-cut separation of normals, carrier parents and their affected children, with only a slight overlap between the first two.

No one doubts that the list of detectable carriers will grow, or that tests will become more discriminating – and, with advances in techniques, cheaper and faster. With enzymes carried in the bloodstream, for example, it should be possible to test for a dozen or more in minutes with analysers already on the market. In short, we could, if we made an effort, do very much more to detect carriers by mass screening of whole populations. The crucial question is whether a full-scale programme of carrier detection would be socially, economically and genetically viable or acceptable.

Biomedical opinion is sharply divided. Some geneticists call carrier screening a low-priority luxury. Nearly all recessive defects are too rare for the enormous effort of mass screening to be worth while. Others urge that screening should be started now. For example, the director of the U.S. National Institutes of Health, Dr James Shannon, has said that genetic counselling and carrier screening is 'becoming not merely a moral obligation of the medical profession but a serious social responsibility as well'.*

To take the social question first, I do not believe, nor do many biologists, that a mass-screening programme would intrude too much on our freedoms and 'rights'. As a start, the programme might screen all married couples, the aim being to see not whether they have recessive genes (we all have them) but whether they have the same recessives. It would not find many. For any given recessive there are twice as many 'at-risk' couples as there are babies born with the defect (this assumes that on average every couple has two children). Even with the commonest serious recessive – cystic fibrosis – this would mean only 1 in every 1,000

---

* Quoted from his introduction to a most detailed, comprehensive review of screening possibilities, in 'Hearings before a Subcommittee of the Committee of Appropriations: Part 4', Department of Health, Education and Welfare (U.S. Government Printing Office, Washington, 1966).

couples; with phenylketonuria a mere 1 in 5,000. These couples could then be warned, and warned before they had the affected child that is the usual danger signal today, of their 1 in 4 risk of affected children. When the severity of the disease or the rigours of treatment was explained to them, it seems likely from present genetic counselling experience that most would voluntarily refrain from parenthood. Instead they could adopt or opt for artificial insemination by a donor who was cleared for that recessive. If there was a contraceptive failure and the couple had a natural pregnancy there would be a strong case for abortion (the risks of defect are about the same as for a pregnant woman who gets German measles).

The main problems, though, are how voluntary or compulsory mass screening should be – we are dealing with healthy people – and what pressures society should put (if any) on the at-risk couples that screening identifies. From the popular demand for cervical cancer screening, it looks as though compulsion would be far from necessary. In fact the main complaint might be that screening is being developed too slowly. If screening were universal and detected couples had no natural children – in other words, if there were maximum compulsion – the defect would be almost completely eliminated. This would be a large gain – for families and society – but the cost to liberty might be too high.

The best way round this dilemma is to screen people before they are married; or, better still, before they are old enough even to be at risk of meeting someone they might want to marry; or, best of all, when they are born. The further back one goes the less obtrusive screening becomes. The tests themselves – even if compulsory – would be progressively easier to accept and arrange. Think of the difference between tests for the married, for couples when they announce their engagement, captive populations of schoolchildren as part of a routine school health programme, and for babies, by taking a couple of spoonfuls of cord blood at birth. But, more important, the consequences of early testing are less obtrusive.

If everyone were tested at birth or school, in principle they could avoid the agony of falling in love with someone and *then* discovering their genetic incompatibility. There is an interesting problem in communications here. If this agony is to be avoided one must know that this new acquaintance is genetically impossible for one as a mate *before* he or she begins to seem emotionally probable! But to ask for this genetic information early in a developing relationship would suggest marital intentions: giving one's recessive list to a girl at a party would make a big hole in the conversation. So the signals would have to be visual – noticeable, but discreetly so. Professor Linus Pauling has suggested small tattoos on the forehead, which might seem bizarre, but we must remember that we carry avoidance signals already, including skin colour, language, wealth or poverty, accent, religion and education. Genetic signals in addition might be hardly noticeable, especially since they would say 'stop' only very rarely.

For a disease like phenylketonuria, where 1 in 10,000 children are affected, about 1 in every 50 people are carriers. So the warnings not to get too entangled would be given only once in every 49 meetings between different men and women. For this defect one could marry any one of 49 people, but would be ill-advised to marry the fiftieth. This does not reduce the available fish in the pool by much, and because it doesn't, it also suggests that a society could ban at-risk marriages without too much fuss. There might be outrage at first – with echoes of compulsory vaccination and fluoridation – but it might not be long before it died down. Bans on marriage between relatives is a very old long-accepted tradition in all cultures; extending these consanguinity bans to include con-recessive bans is only a small step. As Professor Peter Medawar has written* (though with the postscript that choice should be voluntary): 'If anyone thinks or has ever thought that religion, wealth, or colour are matters that may properly be taken into account when deciding whether or not a certain marriage is a suitable one, then let him not dare suggest that the genetic welfare of human beings not be given equal weight.'

* Science and the Sanctity of Life', *Encounter* (December 1966), pp. 96–104.

The second question concerns economics. No state is going to start an expensive mass-screening programme unless there are reasonable benefits – which means, broadly, that the cost per test must be low, or the disease common, or the treatment or care costs (plus the 'compassion factor') high. As we have seen, so far this winning combination has only appeared once forcibly enough to lead to mass screening – in Italy, with Cooley's anaemia, which is so common that in some areas about a fifth of the population are carriers. The trend, though, is that it will appear more in future. We had all better prepare ourselves for a furious public debate on the cost-benefit of genetic screening.

The basic arithmetic for it is fairly simple. Suppose that a carrier test for any one disease costs x. Then to screen 1,000,000 people would cost 1,000,000x. With a rare disease (say 1 baby in every 20,000) *untested* 1,000,000 people would produce about 1,000,000 babies of whom 50 would be affected. So if the cost of treating the disease is more than 20,000x, on cold economics alone mass screening would be worth while. With a more common disease like cystic fibrosis (1 in 2,000 babies) there would be 500 affected babies without screening, and therefore it becomes economic if treatment costs more than 2,000x.

On this basis, several diseases appear to be very strong screening candidates. Take cystic fibrosis, for example. It costs several thousand pounds to treat a patient through childhood, so that if x is less than two to five pounds, screening would be economic. Phenylketonuria (P.K.U.) is another likely candidate. The special artificial diet that can now prevent the P.K.U. baby becoming mentally retarded costs (in Britain) at least eleven pounds a week at the time of writing; and as we shall see in Chapter 7 it is often extremely difficult to keep a child on the diet. Since it probably has to be kept to rigorously for at least twenty years the 'cure' costs are a minimum of ten thousand pounds per victim. P.K.U. is now thought to strike 1 in every 10,000 babies, so that the break-even figure for x is about one pound per test. A simple blood test for P.K.U. carriers is almost certainly just round the corner.

In fact a realistic figure for x is *very* much less than one pound.

Machines are now coming on to the market which can analyse blood samples for a dozen or so chemicals at a time, and which work about sixty times faster than a laboratory technician. One of them, for instance – the Vickers multi-channel 300 – can analyse up to three hundred samples an hour, testing each sample for twelve factors, more or less round the clock. Working only half-time for a year it could deal with one and a quarter million samples – if the formidable administrative problems of keeping it supplied and distributing the results could be met. With ancillary equipment, its capital cost is roughly sixty thousand pounds, and it is estimated that if worked fairly hard it can analyse a sample for five pence, including maintenance, chemical reagents, staff costs and so on. But since one sample could give twelve carrier tests, x comes to less than one *penny*. However this sum is adjusted for administrative costs, it appears that once there are blood tests that do not give too many false negatives or positives (and there are several already), carrier testing is definitely economic. Administrative and logistic problems are another matter, but they will be solved – auto-analysers are too important for all branches of preventive medicine for them not to be.

So much for the social and economic questions. What about the genetic ones? These include some of the most fascinating and important problems of all, for once a society starts on a screening policy it is tampering with its own biological evolution in a profound and perhaps irreversible way. And the consequences could be dangerous.

The main danger is this. If we prevent the births of children with recessive defects we dam up the natural leak in the pool of genes for that disease. Normally this pool is kept at a fairly steady level. There is a trickle of 'bad' genes entering it from new mutations, and this more or less exactly equals the outflow of 'bad' genes – the genes that disappear when a victim dies young or survives to adulthood too incapacitated or subfertile to have children. Stop the leak by preventing the birth of individuals with a double dose of the gene, and you stop those two genes leaving the pool. But you do not – you cannot – stop the parent carriers from reproducing: we *all* carry bad recessives.

What a screening programme does is to ensure that a carrier will marry a person normal for that gene so that if they have two children on average one of them will also be a carrier. In other words, the next generation will have the same proportion of carriers. Meanwhile, though, there is that steady trickle of mutations, so the number of carriers will rise. This rise is very slow. With phenylketonuria, for instance, it would be about 1 per cent each generation or, say, 4 per cent each century. Nevertheless, in the very long run most people would be carriers for most screened genes and they would *have* to have children by each other if they were to have children at all. The pool of 'bad' genes would have risen so high that our descendants would be forced to let the preventive dam burst and then check the flood of defects by ever more heroic and costly 'cure' programmes.

In effect, this is very similar to another trend that is already occurring. By improving medical care, more children with genetic defects who would have died in the past are surviving and having children. Given his insulin jabs, the congenital diabetic can now father diabetics. No one could contemplate a policy of banning insulin, so that the only ways to stop this rise are either to ban saved diabetics from having children or to improve our 'cure power' so much that being a diabetic (or whatever) becomes a relatively trivial matter.

What biologists think about this depends on how optimistic or pessimistic they are. Everyone agrees that the 'cure power' of our descendants will be vastly greater than ours – which is certainly increasing much faster than 1 per cent a generation. What no one can agree on – because we simply cannot know – is whether it will get more successful fast enough to meet the challenge of genetic deterioration.

The pessimists believe it will not. They see a huge escalation in the cost of curing, so that more and more people spend their lives keeping the rest going. They also see the cures becoming increasingly aggressive and life-limiting: 'It takes little effort to conjure up glimpses of the bizarre brave new world – a world of enormous individual variability, each individual (human?) uni-quely wired up and supported by his own special set of transplants and external biochemical plant ... An individual that would be recognizable as a member of *Homo sapiens* would be rare indeed.'* The pessimists' conclusion is that defects must be accepted, the genetic *status quo* must be maintained and evolution must be left untouched, for any attempt to buy jam today is bound to cost our descendants dear.

At the other extreme, the biological engineering enthusiasts believe that cures for defects will become cheaper and less demanding. Instead of the daily insulin injection, the diabetic will be fitted with a neat, long-acting insulin capsule or a transplant of normal pancreas cells which correct the insulin deficiency. And so for all the other biochemical-genetic defects. In short, the wizardry of tomorrow's medicine will make negative eugenics increasingly unnecessary.

What happens, of course, will be somewhere in between. Where, we do not know till we get there. Because of this, almost inevitably, we will act to get the maximum jam *today* – and let the future look after itself. This is how we have always acted on technical innovations and it is inconceivable that possible difficulties centuries hence will alter what we do now – especially when what we are doing now is trying to give individuals the chance of preventing one of *their* children from being a genetic cripple.

This may be 'right' or 'wrong'; we shall have to talk about it. But whatever we do decide on, there is nothing immutable about our choice. Our descendants can always get off the down escalator of genetic 'deterioration' by revising their screening policies. They would just have to ask themselves, as we ask now, whether a defect was so 'curable' that it was not worth trying to prevent – as we now judge the astigmatic or diabetic not worth trying to cure because we can provide him with glasses or insulin. Our descendants, with fuller evidence, will have the privilege of changing their minds.

It is possible, if molecular biologists can break down some

*Leonard Ornstein, 'The Population Explosion, "Conservative Eugenics", and Human Evolution', *Bulletin of the Atomic Scientist* (June 1967), pp. 57–60.

stiff technical barriers, that they will also have the privilege of being able to change their genes.

## Genetic engineering

Five years ago most biologists were confidently predicting that genetic engineering or gene surgery was a very remote prospect – something for the twenty-first rather than the twentieth century. It probably still is if one thinks in terms of its widespread, routine use. But genetics has moved so far, so fast, in the past few years that only the most conservative biologists now believe that human genetic engineering won't be something to reckon with in the next 10–15 years, or possibly even sooner. Indeed, as we shall see, the first tentative steps in applying genetic engineering techniques to human patients have already been made. It is also significant that in 1970 the U.S. Congress voted $10 million to fund a Genetic Task Force whose principal aim is to advance and *apply* genetic engineering methods.

To see what is involved, one has to start by looking at the raw material the aspiring gene surgeon or engineer has to work on. His ultimate task is to edit the master tapes of life – the wispy, thread-like molecules of DNA in every cell which carry the hereditary message – and to edit them so precisely and controllably that a single defective gene here or there can be snipped out and replaced by a normal one. His first problem is that these tapes, his 'patients', are small. In each human cell there is about one yard of DNA, but the DNA molecules are so thin (two-millionths of a centimetre) that together they weigh only six-millionths of a millionth of a gram. If one put together the DNA from all the fertilized eggs from which all the present 3,600 million members of the human race began it would add up to a pinch of matter weighing one-fiftieth of a gram (or a tenth as much as a postage stamp). An individual gene, of course, is far smaller still. No one knows how many human genes there are, but guesses range from fifty thousand to one million – all of them packed on to those three feet of DNA tape.

Second, DNA is an almost infinitely monotonous molecule.

Stripped to its barest essentials, it is a framework for holding a long string of four different chemical groups, called base-pairs. These are the four symbols of the genetic code which the cell can 'read' and so get instructions to put the building blocks for proteins into precisely the right order. In symbol form, then, a stretch of DNA might look like this:

... AGACGTCTTAAGCGTAGCGACCTATATTGCCCGAGTA GGACCTATGCGACTAGCGCAAACACCATCGCTGA ...

A length of this message long enough to code for a protein is a gene, and the code works in such a way that three symbols are needed for each protein building block. Thus a gene for a protein with 100 building blocks would be a stretch of DNA 300 base-pairs long with a few extras on the end for telling the cell to start and stop reading the code.

In most genetic defects, even the most crippling ones, the 'only' fault is that just one protein building block is wrong. In other words, a group of three symbols at a particular point on the DNA tape for that protein is 'wrong: for example, instead of ... ACG ... there might be ... CCG ... The genetic surgeon's task is to change the first 'C' into an 'A'. Since there are about 1,000 million base-pairs in the complete genetic message, and so roughly a quarter that number of 'C's that must *not* be changed into 'A's, he has a tricky job to do. It has been compared to altering a single letter in a copy of the Bible which cannot be opened and which has been shrunk to the size of a pin-head. To make things worse, one might add that no one has read more than a paragraph or two of the hereditary Bible. The positions of only a few human genes have been mapped in the chromosomes – and those only crudely – while even if one could locate any gene exactly in the tangled mass of DNA tapes, biologists only know the base-pair sequence of a very, very few of them.

One might ask why anyone believes gene surgery could ever be possible. Well, at this precise, ultra-fine level it probably never will be: the idea of changing single base-pairs by knocking atoms off with ultra-fine radiation beams, or treating DNA with highly selective chemicals, is almost certainly for science fiction only.

However, if you lace the current explosive progress of genetics with a dash of optimism, several other though cruder ways of altering genes could be used.

The main approach that biologists are exploring is to use *living* surgical tools to stitch in and snip out whole genes or gene fragments. One method depends on a process called transformation, in which bacteria can be made to infect other bacterial cells with their genes. In another process called transduction, viruses can pick up genes from one cell, transfer them to another and deposit them there. In both cases the foreign genes become a functional part of the host cell's genetic apparatus.

Scores of research groups have already used these and similar tools to do some surprisingly sophisticated genetic manipulations on plant and bacterial cells. Lengths of DNA in the host cells have been altered or deleted, and new 'foreign' lengths added, pointing to the possibility of modifying existing faulty genes in human egg cells so that they work normally. Whole genes have been transferred from cell to cell, suggesting that gene transplants might be possible: a carefully selected team of viruses would enter a normal cell culture, pick up the genes of a particular type (the haemoglobin genes, say), transfer these to the recipient cell culture, delete the abnormal haemoglobin genes there, and replace them with the normal ones. Ultimately it might be possible to use similar techniques to introduce artificially synthesized genes: in 1970 a research team at the University of Wisconsin led by Dr H. G. Khorana announced that they had totally synthesized the first complete gene, a stretch of DNA 77 base pairs long which controls the production of the enzyme called 'alanine transfer RNA' in a certain virus. Some human genes, such as the one for insulin, are only twice this length, so the prospect of synthesizing simpler human genes is not all that far-fetched.

In other experiments, inactive genes in cells have been 'switched on' so that they start to produce enzymes that the cell normally doesn't make, suggesting that the human gene surgeon could have a much wider variety of cell types to operate on than was once supposed. Equally important, with plant cells it is now possible to introduce foreign RNA (ribonucleic acid) – an intermediary in the process by which DNA manufactures enzymes – into a cell and thus make the cell produce the particular enzyme controlled by the particular type of RNA injected into it.

Despite these remarkable achievements in gene manipulation, as tools for human genetic engineering they are as yet hopelessly inadequate. The biggest drawback is that viruses and bacteria are extremely inefficient tools for manipulating genetic material: to make a specific gene change in a bacterial cell one normally operates on millions of bacteria in the hope that one or two cells will be altered in the desired way. The gene surgeon will obviously have to have much more precise tools than this when he operates on human eggs or blastocysts.

However, in March 1971 a group under Professor Henry Harris at Oxford University reported an epochal series of experiments that could well open the door to successful human genetic engineering in the not-too-distant future. What they did was to insert into a deficient mouse cell the specific genetic material needed to enable it to function normally.

The deficiency was the inability to manufacture an enzyme called inosinic acid pyrophosphorylase (IAP), a genetic fault which in humans causes the rare but fatal condition called the Lesch-Nyan syndrome. Their technique was to fuse the deficient mouse cells with normal cells from chickens, to make cultures of mouse-chicken hybrids. In this fusion process the new hybrid cells each contain two nuclei but the nuclei start to divide at different times, with the result that the first to divide (in this case the mouse nucleus) takes over and the second nucleus (chicken) is rapidly broken down into fragments.

Harris and his team hoped that during this breakdown any 'useful' bits of the chicken nucleus would be taken up by the mouse nucleus. They found that this is exactly what did happen: with extraordinary precision the deficient mouse nucleus picked up just the gene it needed from the fragments of chicken nucleus.

It is this degree of precision which makes the experiment so striking, and promising, for clinical application to man; the fact that there was no gross transfer of genetic material but only a one-way swap of just the gene that the mouse cells required. Nor is the ability confined to just these species: Harris's group has also transferred genes from toad to mouse cells and from mouse to hamster cells and at the time of writing was planning to try transfers to human cells and to whole animals.

However, as with all other methods of gene transfer, today's methods aren't anything like capable of scoring a success with every single cell. Because of this it is almost certain that the first human genetic engineering will have to be done on a hit-or-miss basis, and therefore on *body* cells, not germ cells. As Dr Edward Tatum, a leading genetic engineering optimist, has described it,[*] 'The first successful genetic engineering will be done with the patient's own cells, for example, liver cells, grown in culture. The desired new gene will be introduced . . . [and] the rare cell with the desired change will then be selected, grown in a mass culture, and reimplanted in the patient's liver.' Re-programming cells in this way could become an important tool for treating genetic defects. For example, by re-programming the liver cells of a phenylketonuric child, one might give him a normal set of liver enzymes that would effect a total, permanent cure for his disease. On the other hand, other treatments might turn out to be far easier and cheaper (see Chapter 7). Many molecular biologists believe that by the time genetic engineering is safe enough to use on man, more straightforward techniques will be developed: the phenylketonuric, for instance, may get a routine liver transplant.

In fact this first type of genetic engineering, on body cells, has already been attempted on human patients due to a peculiar fluke of nature. In 1970 Dr Stanfield Rogers, of the Oak Ridge National Laboratory, Tennessee, heard of two German sisters, aged 5 years and 18 months, who suffered from severe mental retardation, paraplegia and epileptic seizures because they had been born with a rare 'missing gene' defect which meant that

[*] Reported in *Science*, 153 (1966), pp. 442–8.

they could not manufacture the enzyme arginase. As a result they were being 'poisoned' by abnormally high levels of another chemical, arginine, in their blood.

Several years previously, Dr Rogers and his Oak Ridge colleagues had found that when a certain virus – the Shope papilloma virus – infected rabbit cells it induced them to synthesize arginase, presumably by interfering with their genetic machinery. Several laboratory workers who had handled the virus were found to have abnormally low amounts of arginine in their blood (with no ill effects), presumably again because the virus had raised their genetic capacity to make arginase. The Shope virus thus seemed like a possible answer to the German sisters' inability to produce arginase and they were duly infected with it. Unfortunately, either because the dose was insufficient or the vaccine was too impure, little happened. However, the principle of sending in a 'viral surgeon' to alter a human patient's genetic apparatus in a precise way – even though it may have been due to a rare, untypical discovery – does make this event a significant landmark in medical history.

It is a very long way from this type of genetic surgery to true genetic engineering on human *germ* cells rather than body cells. If this could be done with high precision, then not only would all the cells of the resulting person be altered, but so would those of all his descendants. Harmful genes could be truly eliminated. But the difficulties are formidable. Not the least of them is that one would have to be 100 per cent sure of one's technique before daring to doctor the genes of a sperm or ovum or fertilized egg and then re-implanting the 'patient' in a woman to develop into a child – perhaps with monstrous results if one had slipped up.

In the first edition of this book I followed the opinions of most biologists and predicted that this ultimate step in genetic engineering is hardly likely to come in this century. Nowadays one cannot be so complacent, mainly because of remarkable advances in techniques not so much of doctoring genes but of doctoring the earliest stages of embryonic life.

It is now known that up to the moment when it implants itself in the womb as a 100-odd cell blastocyst, the developing mammal egg has a peculiar ability to tolerate various manipulations – and survive, though in an altered form. For instance, if a few cells from a strain of black-haired rats are injected into a blastocyst from an all-white (albino) rat, the blastocyst will grow into a white rat with black patches of skin. The resulting animal is called a genetic mosaic: it is literally a patchwork combination of two (or more) different genetic types. Furthermore, the degree of mosaicism can be varied at will: if, for instance, one foreign cell is introduced at the 16-cell stage of the blastocyst, the end result will be a mosaic animal that is fifteen parts normal, and one part foreign. Similarly, by introducing one foreign cell at the 4-cell stage, one can make a mosaic with one quarter of its cells of the foreign type.

The implications for genetic engineering are clear enough. If a couple is known to be at risk of having children with a genetic defect – say haemophilia – in principle all one has to do is to fertilize their eggs and sperm in the test-tube and innoculate several blastocysts with a fair proportion of normal, non-haemophilic cells, implanting the healthiest survivor in the woman's uterus. The resulting child will be a haemophilic-normal mosaic but – on present understanding, at least – should appear perfectly normal and function normally. With many 'missing gene' defects it is thought that only a slight degree of mosaicism – enough to raise the missing enzyme from zero to about 10 per cent of normal levels – would effectively cure the defect and allow normal chemical functioning.

Do we have anything to fear from all this? Along with the hopeful prospects of eradicating defects, could there be any substance in all those sinister visions of scientists tinkering with human genes in misguided attempts to 'improve' the species or design specialized men for specialized slots in society?

I do not think so for a moment. When one tries to think of specific evils they tend to crumble away. In war, for example, any notion of a 'gene bomb' – perhaps a phial of transducing viruses dropped in an enemy's water supply – already has far quicker-acting and more devastating competitors in today's bio-logical- and chemical-warfare arsenals. If 'improving' men is the evil to fear, general qualities like intelligence are determined by so many genes (and so largely by the post-birth environment also) that tinkering with them effectively is almost inconceivable. Haldane once said, the only problem with creating a race of human angels is to find the genes for wings and for moral perfection: the gene surgeon who wants to make 'better babies' faces a no less daunting task. Special qualities that may turn out to be due to a very few genes, thus offering the possibility of manipulation – fast reflexes, say, or good vision – are hardly going to appeal to the customers, when to get them they will have to procreate from test-tubes, take perhaps a high risk of failure, and pay a fat fee for the privilege.

There can be little doubt that, for a very long time to come, genetic surgery will be highly experimental, risky and pretty ineffective. This means that it will be strictly limited to cases where these disadvantages can be balanced against a real gain: the chance of eliminating from an individual offspring a truly un-pleasant, life-limiting genetic defect. Seen in this light, the uproar that the first real attempts are bound to cause will surely be concerned – as with heart transplants – almost entirely with whether the attempt is premature or not because of the risks.

Like all new technologies, genetic engineering will have un-foreseen and even unforeseeable side-effects, so we cannot be blandly complacent about it. It is probably too early now to start the detailed public debate on the implications of gene surgery, since we have very little idea of how it will actually be done. In the meantime, perhaps the best advice is to keep calm about it. So let the last words of this section, the final message before we move on to the foetus and adult life, come from the American molecular geneticist, Salvador Luria: 'I would say that nothing at present would justify either the prediction of a coming millennium of human betterment or the proclamation of an im-pending danger of genetic enslavement. Geneticists are not ready to conquer the earth, either for good or evil.'*

*T. M. Sonnenborn, ed., *The Control of Human Heredity and Evolution* (Macmillan Co., New York, 1956), p. 16.

# 6. FOETAL MEDICINE

A human egg is so small that something like fifty million can be packed inside the shell of a hen's egg. Yet in nine months, by doubling itself about thirty-five times, each of these tiny scraps of material turns itself into a staggeringly complex but exquisitely well-organized community of cells, with many more members than there are people in the world. It is an astonishing process and an almost totally baffling one. To be blunt, for the first nine months of life we are about as cut off from scientific understanding and medical care as a Neanderthaler in his cave.

But times are changing. Whilst most of our moral and practical attitudes to the foetus are still grounded in a tradition of blissful ignorance of what goes on in the womb, biomedicine is moving in with increasing speed and urgency.

One of its main motives is sheer scientific interest. Now that biologists know in some detail what genes are, the next big questions to answer are about how they and the environment together programme growth and development. But the most compelling reason is practical. Quite simply, the womb has become the most perilous environment in which humans have to live.

A century ago there was a fairly even chance of dying at every age range. But now that most of the relatively easy-to-conquer hazards like infectious diseases have been swept out of sight we are left with the intractable ones at either end of the age span – the degenerative and 'social' diseases of middle and old age at one end, and the pre-natal factors at the other. So nowadays, though some 65 per cent of human beings ever conceived in the developed countries do not die until they are 65 or more, about 20 per cent do die natural deaths before they are ever born. This percentage does not include induced abortion. It also ignores the fact that many post-natal deaths are due to pre-natal causes, and it ignores the heavy toll of handicapped survivors from the womb.

The human uterus is an extraordinarily dangerous environment. Of 1,000 human conceptuses – about a quarter of a night's production in Britain – between 120 and 150 will be dead within four weeks. Mercifully, most of these deaths will be unnoticed, except perhaps as a peculiarly heavy menstrual discharge. But then, between the first and seventh month of pregnancy, there will be a further 100 to 150 deaths. These will be noticed – as spontaneous miscarriages – and they will cause much grief. Usually, though, these deaths are for the best. More than half of all spontaneously aborted foetuses can be seen to be abnormal even from a superficial examination, many of them grossly so. Some experts say it is unusual to find one that is normal. As for the survivors of this tough obstacle course, in the developed countries roughly 2 per cent of babies are born dead and another 3 per cent die before they are fourteen years old. Of these five deaths in every hundred births, four are due to pre-birth or birth causes. On top of this there are the 4 per cent of babies who live with serious or mild handicaps incurred during pregnancy and delivery. Of these, very roughly, threequarters are not due to bad genes but to a wrong uterine environment or a birth accident.

Clearly, we should think ourselves lucky to have got out of the womb alive and well. Equally clearly, the pressure is there to drive the medicine of the foetus and its pregnant host fast and far. So in this chapter, to put it briefly, we look at several major areas where foetal medicine may challenge our moral, social and other attitudes and – equally important – where we have to form our attitudes to help to steer its course into the future.

First, we must find out what causes congenital deaths and defects. As we saw in the last chapter there are genetic causes, environmental causes and unknown causes. The last is by far the biggest group. It includes nearly all the commoner defects, from anencephaly (literally, no head) to cleft palates and hare lips, and it includes all prematurity and most maternal diseases. Obviously, there is little hope of prevention or cure until specific causes are found, and this is going to be one of the priorities of foetal medicine for years to come.

One way epidemiologists would like to trace causes is by prospective rather than retrospective studies. Instead of trying to find whether a defect is caused by a drug, virus, diet, genes or whatever by asking the mothers of affected newborns what they have been doing during the previous year – a notoriously unproductive procedure – they would like to take tens of thousands of couples who intend having children and study them right through their pregnancies. They would need detailed information on such things as diet, exercise, rest and physical and emotional stress, as well as the more obvious factors such as illnesses and drug consumption. Daily or weekly questionnaires would keep memories fresh, making it far more likely that when the few affected babies do appear their parents will reveal a common causative factor. Such a prospective study would have discovered what thalidomide was doing almost as soon as the first thalidomide children appeared. But of course it demands massive effort and cooperation from ordinary, healthy people. But which ordinary, healthy people would be prepared to do that much?

Another approach the epidemiologists would like to see is a vast computerization of their search. They would like to put the complete life histories of an enormous number of families on punched cards, starting with the defect records, age of childbearing, fertility, addresses and so on of one's ancestors, through one's own complete medical, educational and reproductive history, to those of one's children, brothers, sisters, wife or husband. In short, to compile a mass pedigree for any data that might have a bearing on genetic or environmental defects. Computers are becoming so cheap, rapid, accurate and efficient that these streamlined record-linkages schemes are already being seriously discussed. They will involve great effort, expense, huge changes in medical record-keeping and even changes in the law. Perhaps the hardest change to accept is that everyone will have to be provided with a unique number, so that the computers know precisely whom they are dealing with. We shall have to be labelled for the machines. Outside Scandinavia, no country has such a system, yet, without one, mass linkage schemes and even mass computer-aided disease screening programmes are thought to be impossible. Though we may each be on many different sets – we have social security numbers and our names – there is no single, all-embracing set that specifies every individual uniquely. This is the overall problem. The practical-political problem is how to persuade one state agency to accept the chore of expanding its number system to include everyone!

Despite these difficulties, many epidemiologists believe the overall result would be enormously worth while. Some of them also fear that the main problem they will hit is a widespread outcry about invasion of privacy. Will the computers be misused to probe our private secrets? There may be real dangers here: each scheme ought to be thoroughly scrutinized as soon as it is proposed.

A third way of improving the search for causes hits even harder at our traditions and privacies. It asks us to cut through all the mystique and ambivalence about the sanctity and rights of the foetus and to realize that it could be an important experimental 'animal'.

In Britain roughly one thousand spontaneously aborted foetuses die each day. Some are examined, but most are wrapped up and thrown away – either by the attendant medical personnel or by the parents themselves. It might be better to suppress our instinct to be rid of them quickly, keep them and have an expert study them carefully. Each foetus could give a clue about why certain kinds of women, or certain kinds of drug, diet, environment and so on, produce the defects that most of the foetuses will carry. Equally important, they might often act as specific warning signals for individual parents: the discovery of chromosome defects in the foetus could in some cases give a warning that they might appear again – but this time produce a handicapped child.

There do not seem to be any basic religious or moral reasons why we should not do this. No Church baptizes or demands burial rights for a foetus before the seventh month, when traditionally it becomes viable. It is just a matter of balancing the strong human reasons for wanting to discard the dead foetus as quickly as possible against the fact that doing so to some extent guarantees the future births of crippled babies.

Much the same is true of induced abortions. Here too potentially very important experimental material is usually wasted. One exception is in Czechoslovakia, where there is a programme to get women who are going to have an abortion to take various drugs to test their damaging effect on the foetus. After a few weeks of taking the drug the mother has the abortion and the aborted foetus is examined for signs of defect. If this seems like an excess of ruthless efficiency, it is at least a rational and productive approach. Again, it might have nipped the thalidomide tragedy in the bud. On the other hand, there are strong arguments against it. What does one do if a woman changes her mind about the abortion; or rather, how does one select women for the programme who know they will *not* change their minds? And if an abortion is *planned* very early in pregnancy, is it not highly unethical not to perform it then rather than two months later, when the medical risks and psychological impact are intensified?

A greater limit on research is the strong opposition to keeping human embryos alive in the 'test-tube'. This can mean anything from trying to grow a score or so of cells in the blastocyst stage – the point of development normally reached a few days after conception – to trying to keep alive three-month-old spontaneously aborted embryos in artificial placentas. The aim of this work is not to perfect plastic and metal wombs to usher in a brave new world of test-tube babies but to probe, directly, the biochemical and other needs of the foetus in the hope that they will throw light on normal and abnormal *natural* development. All the biologists doing this kind of work that I have talked to say they live with a permanent, nagging threat of exposure by some outraged priest or journalist. More important, they find it hard to get hold of suitable material. To get a living aborted foetus, for example, the scientist must either be a clinician as well or work very closely with one. Most clinicians, though, will not cooperate, either because they would be breaking the law by handing over 'living' human material for experiments or because they would be going against their ethical codes. So as a result very, very few scientists are working directly on human (rather than animal) development.

Of course, there are very real problems here. The most difficult one, perhaps, is how to ensure that the mother of the foetus gives her free and *informed* consent to the experiment. Losing one's foetus is hard enough without knowing that it might be lingering on (though without sensation) as an experimental animal in a test-tube for a few hours. Tradition and humane compassion say that for the mother's sake the dead foetus should be disposed of quickly and decently. They also say that one does not use humans solely as experimental animals. But, given the proper safeguards and consents, can one make exceptions for the pre-viable human foetus – the aborted foetus or the rejected egg that would die anyway? If we continue to say 'no', or even fail to give a clear moral and legal 'yes', research on human defects will be that much limited.

The trouble is that this kind of research has really to be done on humans. Animals are rarely good models for men when it comes to testing the effects of drugs and other teratogens on the foetus. Ideally, one day, teratology will be so advanced that it will be able to explain the links between teratogens and the defects they cause at a biochemical and molecular level, so that what a drug does to a mouse *will* be more accurately translatable to man. When we get that sophistication there will be far less need or pressure to use human material for experiments. But in the mean-time we have to decide what our priorities are.

Another set of problems comes from our growing ability to inspect the foetus in its womb-cave. Biomedicine is opening windows that look into the womb so that we can see better how well – or badly – the foetus is doing.

It is astonishing how small and opaque these windows are at present. Consider, for example, what tests even a good ante-natal clinic does on a pregnant mother during her nine-month wait. There will be a pelvic examination to see if the opening is large enough for a normal delivery, and a blood-group check in case transfusions will be needed. The mother's general health is checked briefly and her blood and urine tested for signs of anaemia, diabetes, syphilis and other diseases. A high weight gain and blood pressure, plus swelling round the ankles, can be early signs of toxaemia – an important cause of maternal and baby death. But the only tests on the foetus itself are usually a brief check on its heart-beat and a bit of prodding about to estimate the stage of pregnancy and whether the foetus is a normal size for its date. If there is any suspicion about it the doctors may make an X-ray examination, but that can tell us little. X-rays can usually detect twins, conjoined twins, severe skeletal disorders like congenital dwarfism or the missing limbs of the thalidomide baby or, if severe, spine and head defects like anencephaly or hydrocephalus. Yet as I write, a friend in her ninth month of pregnancy is still waiting for one of the top London teaching hospitals to decide if her X-rays show whether or not her foetus is upside down.

All this is changing as medical technology develops more discerning foetal probes. For example, there is ultrasonography, a kind of sound radar which sends ultrasonic waves through to the foetus and builds up a picture of it from the way they are reflected back. Unlike radiography, it is totally safe, and already, even in its present early stage of development, can produce clearer and more detailed pictures; for instance, it can detect pregnancy even before a urine test, two weeks after conception.

A more direct probe, though only relevant to late pregnancy, is amnioscopy. Foetal oxygen starvation is the most important single killer of newborns, but until recently there has been no sure way of detecting it. As a result, many births are induced prematurely on suspicion of foetal distress, but as it turns out often unnecessarily. Amnioscopy gives a direct measurement any time in the last six weeks of pregnancy. A tube with a light is inserted through the vagina and cervix to inspect the colour of the amniotic fluid without breaking the membranes keeping it intact. If it detects any meconium – the greasy black stuff that fills the foetal gut – in the fluid it is a fairly good sign that the foetus is approaching oxygen starvation. An even surer method, if delivery is imminent, is to go through the membranes and take samples of foetal blood by nicking the scalp and collecting the blood in a catheter, using the amnioscope as a visual guide. Analysis of the acid and base levels in the blood, which only takes three minutes, gives a very sure assessment of foetal distress. The operation is simple, is said not to increase the risk of infection, and can be done on about 90 per cent of women without anaesthetics. In one of the few centres where it has been used – Queen Charlotte's Hospital, London – Caesarian sections for foetal distress have been halved and perinatal deaths have dropped appreciably.

By far the most dramatic new foetal probe is amniocentesis. It depends on the fact that early in pregnancy the foetus starts to shed some of its own cells into the amniotic fluid surrounding it. By passing a hollow needle through the abdominal wall of the mother it is possible to collect some of these cells in about ten cubic centimetres of the fluid. So one now has something rather remarkable: living fragments of the foetus that can be cultured in the laboratory to provide a kind of parallel model of the foetus – and a model that can be probed and tested as thoroughly as one likes.

For example, with standard techniques, a thorough analysis of the foetal chromosomes can be made. This can reveal with absolute certainty whether the foetus has a chromosome defect such as mongolism – something that was never possible before. With a simple chromosome staining technique one can also determine with near-100-per-cent certainty the sex of the cultured cells, and so of the foetus. This is crucially important with any of the sex-linked genetic diseases, like haemophilia or inherited muscular dystrophy. Only male babies are affected by these diseases, and then only if the bad gene is inherited from a carrier mother. So if a mother is suspected of being a carrier, it is reasonable that she and her husband will only want girl babies and not boys – who have a fifty-fifty chance of being affected.

Ideally they would achieve this by selecting the sex of their children before conception, but determining the sex of a foetus by amniocentisis, followed by abortion if the foetus is male, would be an effective second best. (A better method, now being explored by an American researcher, Dr Landrum Shettles, is to take a swab of cells from the cervix of a pregnant woman. A few of these cells will have been shed by the foetus and can be tested to see if they are male by a simple fluorescent dye which produces a bright spot under the microscope where there is a Y chromosome, indicating that the foetus is male.)

In future other types of test might be possible. One hope is that accurate virus tests can be developed. At present there are reliable tests for whether a pregnant woman has any of the risky virus diseases, but none that can tell whether the foetus is affected, which is what matters. So the question of abortion is entirely based on general estimates of risk. Unfortunately, virus tests on cultured cells have been disappointing up to now. For instance, even foetuses known to have rubella (German measles) damage from tests after they have been aborted have not shown the damage in their shed amniotic cells. But techniques are advancing so rapidly that no one working in this area seems to doubt that direct virus tests on the foetus model will soon be possible.

There are also growing hopes of detecting biochemical abnormalities by amniocentesis, indicating specific gene defects. In April 1971 an American team reported that with 15 pregnant women whose foetuses were known to be at risk from a crippling defect called Tay–Sachs disease, amniocentesis plus tests of the foetal cells for the presence of the 'Tay Sachs enzyme' had shown in every case with 100 per cent success, whether the foetus did or did not have the defect.

Tests like this are still experimental, expensive and slightly dangerous. It will be a long time before they become a routine part of ante-natal medicine, if they ever do. There is also the serious limitation at present that they are not reliable until about the fifteenth week of pregnancy.*

* Only then does the foetus shed enough cells for one to be able to collect, culture and test a statistically valid sample. Not many people would condemn a foetus on finding the mongol chromosome error in one cell of its culture model. The finding might itself be due to a technical error; even if it was not, the foetus might be a mongol mosaic with only a quarter or an eighth of its cells bearing wrong chromosomes. If so, it would become a very nearly normal child. Similarly with sex detection: the earliest that 100 per cent reliable tests can be done is around five months.

Even so, it is not too early to start thinking about the shattering effect this kind of detection power will have on our traditional, ignorance-based attitudes to the foetus – a tradition that certainly does not include the embarrassment of being able to predict with certainty a monstrous birth.

One obvious effect is that abortions for foetal indications will become less of a gamble. Nowadays, wherever such abortions are legal (and often where they are not) most doctors will abort if the defect is severe and the risk is higher than the 20 to 30 per cent usually cited for a mother known to have had rubella in the first three months. They would rather take the higher chance of killing a normal foetus than the low one of letting an abnormal foetus go to term against the mother's wish. Some, however, would not. Early detection will help to cut through this fog of uncertainty. It will make it almost impossible to refuse abortion when the defect is diagnosed for certain; at the same time, it will reduce the number of foetuses that are aborted but turn out to have been normal.

A second obvious effect will be to force many countries to legalize foetal indication abortions. In some of these countries the laws are astonishingly perverse. In many states of the U.S.A., for instance, an abortion in the first three months of pregnancy still carries a possible (though rarely enforced) sentence of from fifteen to twenty years' imprisonment. But after four and a half months – when amniocentesis becomes possible and even an old-fashioned X-ray can begin to detect an acranial or microcephalic monster – abortion becomes manslaughter, with a maximum sentence of ninety-nine years. In other words, the sanctity-of-life ethic, which gets stronger with foetal age, runs totally contrary to the ethic that foetal detection suggests: namely, do not abort until one is sure of a defect, which means later rather than sooner. Yet this ethic in turn leads to the equally perverse conclusion that the best time to 'abort' is when one is surest of all: namely, after birth.

This brings us to the third and perhaps most difficult effect to cope with. What happens when foetal detection picks up sure indications of only moderately severe defect? Most doctors would agree with most parents that abortion followed by a normal baby would be better than *knowingly* producing a mongol. But what if the diagnosis is, say, a sex-chromosome abnormality that leads to slight mental deficiency or sexual abnormalities? Many parents would still want an abortion, but they might find one very hard to get. It is not just that they would have the whole 'save, salvage and care' ethic against them. They face the special problem that the abortion would have to be a late one. When the amniocentesis results come through – say in the sixteenth to eighteenth week of pregnancy at the very earliest – it will be far too late for a simple surgical abortion. Instead, abortion would have to be by saline injection (which pickles the foetus to death before it is expelled normally) or by a Caesarian. Both are particularly nasty to do – especially after the twentieth week, when the foetus is all too lifelike and, indeed, may live for a short time after the abortion. Many doctors will not do them except in extreme hardship cases where the mother's life is risked if pregnancy continues, and it is a well-known trick with 'social' abortions for doctors to prevaricate until after the surgical limit of about twelve weeks and then turn to the mother with an apologetic, 'Sorry, dear, it's too late.'

So foetal detection presents us with a unique dilemma. It gives us certainty that a human being will become subnormal but it does not help us define what is too subnormal to tolerate – what kind of human is not worth producing. And it does this at a time, before birth, when the growing consensus is that the 'human' is not yet fully human and can be killed. Who can tell how medicine, society and individual parents will adapt to this new, uncomfortable knowledge? One can only predict some agonizing doubts and arguments ahead.

There is a fascinating legal twist to this. Recently in the U.S.A. there has been a growing number of 'womb risk' lawsuits where handicapped children have sued their mothers' doctors for refusing to abort them – the plaintiffs – although the doctors knew that the mothers ran a special risk of carrying defective foetuses because of X-ray exposure, German measles, etc. I leave the reader to ponder on the astounding phenomenon of somebody suing somebody else for not preventing him from being born.

The next major impact area is the growth of foetal medicine, or embryatrics. As we learn to detect that a foetus has something wrong with it there are going to be greater pressures to cure it in the womb – at least in all the cases where waiting until after birth is more dangerous.

The only real 'success' story in foetal surgery so far, as most people know, is in treating Rhesus blood disease. Some foetuses have Rhesus-positive blood but a mother who is Rhesus negative. If a few blood cells from the foetus get into the mother's blood-stream through the placenta during labour, the mother produces antibodies against them. During a *later* pregnancy these anti-bodies will cross the placenta to the foetus and, if it is again Rhesus-positive, will attack its blood cells. The foetus gets jaundice and severe anaemia, and may well die shortly before or after birth. The treatment used to be to induce premature labour and transfuse the baby with Rhesus-negative blood. In 1963–4 Dr William Liley of New Zealand realized that amniocentesis, which was already being used to detect Rhesus-risk babies, could be the basis of a cure. With two tubes connected to the blood-stream, the foetal Rhesus-positive blood could be replaced with Rhesus-negative (like the mother's) before birth. The maternal anti-positive antibodies could then do no damage.

Since then, Rhesus-exchange transfusions have saved countless lives and joined the ranks of all those other miracles of modern medicine in popular mythology. What is not so well known is that the transfusion only protects the foetus for a couple of weeks and that roughly 60 per cent of transfused foetuses still die even in the best centres. There are hopes of improving this, though. One possibility now being explored is not to transfuse at all but to inject into the foetal abdomen large doses of liver cells taken from a Rhesus-negative foetus that has been aborted. These will go on making Rhesus-negative blood cells in the recipient foetus for about ten to twelve weeks. There is no rejection problem since a foetus can accept foreign tissue quite happily. The other approach is to prevent the antibodies ever crossing into the foetus in the first place. Pioneered by Professor A. C. Clarke of Liverpool, the idea is to destroy the blood cells that get into the mother from her *first* Rhesus-positive baby by giving her Rhesus-positive antibodies soon after delivery. If the cells are destroyed soon enough the mother never makes her own long-lasting antibodies against them, and so cannot pass them through to any later foetuses. The first trials, reported in 1966, were resoundingly successful. Not one of seventy-eight treated women had any detectable positive cells up to six months after delivery and none of the eighteen who later had further Rhesus-positive babies had any trouble with them. The first ambitions of foetal surgery, in other words, seem to be concerned with a disease that is destined for the museum shelf.

But other ideas are developing. For example, in 1967 Dr J. A. Haller, Jr, reported to the American Association for Thoracic Surgery on the repair of heart defects in dog foetuses – open-heart and open-womb surgery. The foetuses were removed from the mother by a Caesarian, opened up, operated on, closed up and then returned to the mother to recover until they were born. Dr Haller predicted that this must soon be done in man: 'The more familiar we become with the foetal cardiovascular system, the more satisfactory foetal surgery can be. Since we are paediatric surgeons, our goal *of course* is to find ways to salvage human infants with congenital cardiovascular defects. At present too many are lost immediately after birth because we are not getting them early enough.'

The italics are mine. Foetal surgery is only in the foetal stage itself, and no one can predict how it will turn out when it really begins to grow. Perhaps embryatricians will be able to perfect techniques for taking out a foetus, curing its defects completely, and returning it to await a normal birth as a *normal* baby. There is the trauma for the mother and the cost to consider, but no mothers or societies would have serious objections about that. What is worrying is that foetal surgery will probably spread to salvaging foetuses – perhaps earlier and earlier in pregnancy – that might be better left unsalvaged and aborted instead. As we shall see in the next chapter, paediatric medicine is mounting an

increasingly successful, but also increasingly *partially* successful,
attack on defects in infancy. The result is that more and more
children who would have died or been mercifully killed are being
saved – but with residual defects, often very severe ones. When
the alternative is infanticide, this salvage policy is – by tradition at
least – more defensible than when the alternative is abortion, even
abortion in the last two months of pregnancy. Foetal surgery, like
those premature units where keeping a five-month foetus alive is
no longer exceptional, will blur the traditional dividing-line between the foetus and the baby, the invisible 'human' and the
visible human with full human rights. And it will tend to push it
back towards conception. As long as it produces children with
residual defects – or even *risks doing so* – and as long as abortion
for defect is an acceptable alternative, I think we should be very
wary of that.

There has been a lot of speculation that the most extraordinary
impact of foetal medicine may turn out to be to do with improving brains. For a start, there is the South African super-baby
business, which started in the late 1950s when Professor O. S.
Heyns of Witwatersrand University invented his fibreglass dome
and plastic suit technique for lifting the pressure off the abdomens
of women in late pregnancy. The original aim of this decompression technique was to give shorter, less painful confinements and
help to prevent toxaemias of pregnancy. However, two years after
it was first used, reports began to come in that decompression
babies were exceptionally bright. In one study of 250 of them, for
instance, their average development quotient was found to be
about 20 to 30 points above normal, while many were behaving
like genius prodigies. Since then the news has spread rapidly, and
decompression clinics or equipment-hiring schemes (twenty-five
to forty pounds a course in Britain) are springing up everywhere.

Recently, critics have started deflating the bubble boom. The
major objection is that a mother who is anxious enough about her
baby's I.Q. to go in for decompression is likely to be bright herself and, even more important, is going to do everything she can
after birth to push what she thinks is her little genius along. So it

was no surprise to the critics when, in March 1968, Dr Renée Liddicoat of the National Institute for Personnel Research, Johannesburg, came out with a survey of 329 children in which the decompression mothers had been matched for I.Q. with untreated mothers, and which found no significant difference between their babies at one, four, and nine months and at three years. Other researchers have failed to find any differences in the blood acid-base balance of decompression babies. This balance is a more or less direct measure of how well the baby's brain is oxygenated. Yet the foundation of the super-baby claim is that decompression improves oxygenation of the mother's, and hence the foetus's blood at a time when the foetus's oxygen demand is beginning to outstrip the mother's ability to meet it.

Whether or not decompression works, it is part of a significant trend in foetal medicine. As we have seen, oxygen starvation is a major killer of babies and also causes a lot of brain damage. Whatever the causes, the foetus has stayed in the womb too long and it ought to be got out. There is a safer incubator for it in the hospital nursery. Until recently doctors could do very little about this. They could predict when a foetus might be running into real difficulties and ought to be delivered, but could do little to detect mild distress. Now they are getting the techniques to do this. Amnioscopy is one. Measuring the level of a steroid called oestriol in the mother's urine is another: if there is less than 12 milligrams over twenty-four hours it indicates that the uterine environment is defective, and if less than 4 milligrams that the death of the foetus is imminent. So far the measures are crude: they can detect the urgent cases and the normal ones but not the in-betweens. But this is likely to improve. We may be moving rapidly towards a wholly new regime of delivery, in which mothers will be carefully monitored – in hospital – for the last week or two of pregnancy and their babies induced as soon as there are the slightest signs of danger. The gap between this and our present haphazard, leave-it-to-nature methods is so vast (and the costs for all would be so enormous) that one can expect it to generate an enormous amount of angry heat.

To get back to brain-improving, some people have speculated

about another even more dramatic idea – altering the foetal brain directly. Animal experiments have shown that the hormones controlling brain- and nerve-cell growth can be tampered with to produce bigger and 'brighter' brains. For example, Dr Stephen Zamenhoff of the University of California, Los Angeles, has injected pregnant rats with growth hormones in the last two or three weeks of pregnancy and produced baby rats with brains weighing up to 30 per cent more than normal. Most of this extra mass turned out to be in the neurone cells in the cortex – the perception and analysing centres of the brain – while tests have conclusively shown that these treated rats are quicker to learn and slower to forget how to deal with mazes. In the laboratory-rat world that means a higher I.Q.

It seems very unlikely that this could be applied to man. Any injections would probably have to be given in the first three months. Would any parent or doctor agree to run the risks? I think not. There could also be a problem over getting the larger brain born. Professor Lederberg has argued that though the evolution of human brain size has been crucially limited by the size of the human pelvis through which the baby's head has to go at birth – and the number of birth injuries shows just how fine this adjustment is – now that human brains have evolved far enough to have invented Caesarian deliveries we can by-pass this evolutionary road block. In other words, we could artificially induce egg-head foetuses and then cut them out. Perhaps we could. But how many mothers and doctors would accept the trauma, the risks and the sheer extraordinariness of this procedure is another matter.

However, there may be a less drastic way of boosting brains artificially. Brain or cortex size is not all that matters as regards intelligence. Dolphins have bigger brains than we and the brain masses of eminent men (estimated from skull impressions after death) fall well within the normal range. What is almost certainly far more important is the number and nature – the richness – of the interconnections between the brain's 10,000 million neurone cells. As a baby's brain grows and works, the number of neurones stays fixed but the interconnections multiply until each cell is wired to perhaps 10,000 neighbours. The driving force for this extraordinary self-wiring process is partly hormonal (i.e. it happens anyway) and partly stimulation or experience. It seems that the more the brain is used the more self-wiring goes on, the more the wires are 'fixed' as useful pathways rather than being re-absorbed, and the thicker the tangled skein becomes.

So while there is a lot to be said for keeping babies busy there might also be a lot in boosting the foetus's nerve-growth factors in a kind of pre-birth chemical schooling by injection. Scientists are in fact rapidly learning what chemicals they might inject. For example, in February 1968 a group of researchers reported in *Science*\* that they had found a new nerve-growth factor so potent that ten molecules of it can promote the growth of 100 neurone interconnections (in chickens). It is about a million times more active than any previously known growth factor and can be obtained in active yields 10 million times greater than before. What the chickens turned out like has not yet been reported (producing an intelligent hen surely warrants a Nobel Prize), and of course no one knows whether it would work in man without nasty side-effects, or even what the optimum dose would be. Indeed, we might never know. How does one do the first human experiment? And how does one test its effects except by dissecting the brains of the few human guinea pigs that die naturally very soon after the experiment? Yet the possibility is there, just. And so is the far more challenging prospect of a massive dispute between the 'inject them into genius' school, the 'environmental stimulus' approach and, indeed, the 'leave them be, let them enjoy themselves and to hell with forced learning' approach. But more on this in Chapter 8.

While we are in science-fiction land, what of the artificial womb? Is this another major problem we shall have to face? Are we, as so many people seem to believe, going to wake up one morning to read of the first human being who never lived in its mother and then – oh, Brave New World! – will we have to tussle with the awesome problems this will bring even if one assumes

\* I. Schenkein, *et al.*, *Science*, 159 (1968), pp. 640–43.

that the motives for rearing motherless babies are entirely benign?

A short answer is 'no'. In the 1930s, when Aldous Huxley wrote *Brave New World*, it did look as though it might soon be possible. Biologists had just started keeping animal fertilized eggs alive in the test-tube for a few days, while at the other end of pregnancy, incubators were getting better at saving premature babies. It seemed as though extending the two approaches to meet in the middle would be just a matter of time, and not very much time at that. Since then, biologists working in these fields have learned better. They have realized that the astonishingly complex interchange of chemicals between a mother and foetus – an interchange that even millions of years of evolutionary design have not made safe – is not going to be easy to copy, to say the least. For a long time to come, perhaps for ever, attempts to imitate the real thing are going to be very inadequate and very, very unsafe.

Consider the problems. In the most traumatic physiological upheaval we ever live through, birth drags a foetus from a liquid into a gaseous environment. One approach to an artificial womb is to extend the gaseous environment back, as in an incubator. But there is an obvious limit to this, because one reaches a point where the foetal lungs are too underdeveloped to cope. With humans this point – the backward limit of the incubator – is about six months, though occasionally foetuses have survived in incubators from five months (when they weigh about one and a half pounds). But most of these early incubated babies either cannot adapt or do not adapt well: they die, or they live with some loss of (usually mental) function.

To get over this, several biologists have tried using liquid incubators. For example, Dr John Callaghan of the University of Alberta has brought premature lambs to term after removing them from their mothers by a Caesarian and putting them in a plastic bubble filled with sugar and salt solution before they have time to draw a breath and hence irrevocably switch themselves to an air-breathing system. The lamb gets its oxygen in the normal way, through its umbilical cord, which is attached to a heart-lung machine. So far no lambs have lived more than a few days like this, mainly because heart-lung machines damage blood cells after four or five hours. (The aim of this work is in fact to perfect heart-lung machines, not liquid incubators.) But even when gentler heart-lung machines are perfected, as most experts predict they will be quite soon, there is the nasty problem of how one provides the foetus with the correct foods, hormones and all the rest of the myriad chemicals that normally pass from the mother's bloodstream to the foetus through the placenta and umbilical cord.

To provide these one either has to push fresh blood into the system at just the rate that the foetus uses up what the blood contains, or one has to dose the system with these chemicals, as if one were spicing a sauce. The first approach is impracticable for long – where would all the donors of just the right type come from? As for the second, we are just nowhere near knowing what the right recipe is. Until we do know – precisely – the foetus is bound before long to be poisoned to death or serious malfunctioning. These devices are now roughly at the stage where they can keep a near-term animal foetus going for several days and bring it to term and a normal life, or keep a three-to-six-month aborted human embryo alive, but never for more than about forty-eight hours or so. And they have been at this stage, without improving significantly, for several years.

At the beginning of development there are even more difficult problems. Several biologists have grown mammalian and human fertilized eggs in the test-tube. Others have taken animal embryos at later stages of development (equivalent to about fourteen days in man) when organs have just begun to form and tried growing them artificially. So far, despite years of trying, no one has managed to get any of these organisms to survive for more than a few days.* The main reason for this failure is, simply, oxygen

*In April 1971 Dr Y-C Hsu, of Johns Hopkins University, Baltimore, reported a record survival time for mammal embryos of about 14 to 16 days. Starting with blastocysts of mice flushed from their mothers a few hours after fertilization, Dr Hsu grew a small proportion in the test-tube to a point where blood vessels and blood cells developed, and a primitive

starvation. In most of the early experiments the tiny eggs or micro-embryos were planted out in drops of blood serum, from which they could draw oxygen by diffusion through the walls of the outer cells. Being a static system, within a day or two all the oxygen in the serum was used up and the organism died. More recently, some experimenters have tried boosting the oxygen (and nutrient) supply by circulating serum past the cultures. This helps, but not for long. As the organism grows its bulk increases much faster than its surface area, and diffusion becomes inadequate. At some point the embryo has to switch to getting its supplies through a placenta and bloodstream, but artificially grown embryos very rarely produce a viable placenta. Even if this could be overcome there are still some other monumental hurdles to jump. Even supposing some biologist did learn how to rear an egg so that it developed into an embryo and placenta, and learned how to join the minute blood vessels of this living system on to the plastic tubes of an artificial womb, it would be even more crucially important to get the blood chemicals in the machine just right. If there is one thing biomedicine has learned about the embryo in the last twenty years it is its extreme and usually disastrous sensitivity to a long list of chemical agents, or their lack, in early development when organs are forming.

No, for the foreseeable future the full egg-to-baby artificial womb looks impossible. Every biologist working in this field thinks it will be technically impossible for many decades (some think for ever), while the only thing that could make it possible is some unimaginably novel advance in knowledge and technique. Meanwhile the idea that an artificial womb should be used on man before it is *totally* possible is surely unacceptable. Until a machine can copy a natural womb exactly, using it to grow humans would be to run an almost certain risk of deliberately and artificially producing cripples.

---

heart beat had started. Development never went further than that and the embryos died. Significantly, all these comparatively advanced test-tube embryos were grossly defective in one way or another.

Now let us get back to earth. The next and last group of problems we are to look at is far more mundane but far more important than any far-flung speculation. It is that we are gathering knowledge about foetal defects too rapidly to spread it around, absorb it and apply it, while at the same time we have hardly begun to use the knowledge we already have. This is no new problem, of course – in medicine or anything else. But it is peculiarly relevant to the prevention of congenital deaths and defects, and the experts are already debating it hotly.

Very broadly there are two opposite prevention strategies we could adopt. The first is not to bother so much with the new, with research, with trying to probe into the causes of individual defects. The significant fact is that very many foetuses die, or survive with defects, because their mothers are poor or ignorant or both; they are badly housed, or underfed or overcrowded; and above all they have no access to good ante-natal or obstetric care. So the challenge is to make a broad frontal assault by improving living standards and health services.

There is a vast amount of evidence to back this attitude. To give just a few examples, in Britain the best way to get oneself born alive is to be conceived in a tall mother living in the prosperous South of England married to a man in Social Class I or II. The perinatal mortality (still-births plus death within a week) for this group is *one third* what it is for babies whose mothers are short, live in the industrial Midlands and are married into Social Class V. In America the best advice to a foetus who wants to live is not to be conceived in a Negro. Much the same goes for mental subnormality. It has long been known that severe subnormality is no respecter of social class: an idiot is as likely to be born to a duke as to a dustman because the duke and his duchess are as likely to have genetic or chromosomal abnormalities or birth difficulties. But there is now much evidence suggesting that mild subnormality (I.Q. 50–75) without actual organic defects is virtually confined to low-income groups and that though this is partly due to genes and the post-birth environment much of it is due to deprivations during pregnancy and the failure to use medical services.* Prematurity and dysmaturity ('small-for-date' babies) – both major causes of mental subnormality and other defects – are also commoner in lower-income groups. As an Aberdeen study of mentally retarded children has put it, 'Low I.Q., and domestic and occupational disability, are associated with low standards of living and poor nutrition, which lead to stunted growth, poor obstetric performance and low birth weight.'† Such children are neither conceived nor gestated nor born equal: they grow in a poor womb environment, have the poorest labour care and the poorest environment in which to grow – and so they tend to repeat the dismal cycle of *socially transmitted* defect.

More surprisingly, perhaps, the same picture holds for many specific defects. For instance with anencephaly – an invariably lethal failure of the skull to form properly – the rates for Scotland in the 1950s ranged from 1 per 1,000 births in Social Class I to nearly 4 per 1,000 in Social Class V. About the only defects where the poor do not come off worse than or equal to the rich are those caused by viruses. For example, with the cytomegalic virus, which hardly affects adults but can produce babies with spasticity, seizures, deafness and I.Q.s below 35 when it hits mothers early in pregnancy, about 80 per cent of low-income women have caught it before child-bearing age (and so cannot catch it again while pregnant) compared to 50 per cent for middle- and high-income women. The virus is spread by saliva, and the rich, though they kiss, are less often in places where strangers spit.

At birth another wedge is driven between the babies of the rich and the poor. In fact for everyone this is an astonishingly under-developed bit of medical territory. The American obstetrician, Professor Allan C. Barnes, has pointed out that in the eight-week span before and after birth 3·5 per cent of the population

* See for example the chapters by James Walker and Albert Kushlick in J. E. Meade and A. S. Parkes, ed.,*Genetic and Environmental Factors in Human Ability* (Oliver and Boyd, Edinburgh and London, 1966). They include many references to this and similar topics.

† Cited in J. Walker, op. cit.

die and an equal number are damaged or found to be damaged. Much of this is at present unpreventable, but as Barnes, like many experts, says, much could be prevented if doctors and the public took birth more seriously instead of thinking of it as a time of sweetness and light and something that nature has equipped mothers to get on with by themselves. Even in hospitals the birth of most babies only merits one doctor, a couple of nurses and a minimum of equipment, even though the combined life expectation of mother and child is well over a century. Compare that to a brain or heart operation where a score of skilled people and tens of thousands of pounds-worth of gleaming gadgets may be involved, though the patient's expectation of life is usually a few years at best. And most babies, of course, are not born in hospital – especially most under-privileged babies. They are born at home, far from the skills and machines that might save them or their mothers if disaster strikes.

Thomas McKeown, Professor of Social Medicine at Birmingham University, has estimated that if there were better ante-natal and delivery services in Britain – especially more hospital deliveries – then it should be possible to prevent two-thirds of all stillbirths, from a half to two-thirds of infant deaths and probably a considerable amount of long-term defect.

Providing all these services is not just a question of money and manpower. A huge change is also needed in the attitudes of the public as well as of doctors. For example, in a 1966 survey* by the British Royal College of Midwives of 1,514 women who had just had their first babies, 62 per cent said they had feared having an abnormal baby. Yet only 8 per cent of these worriers said they had received help from ante-natal classes and clinics. Many were afraid of making fools of themselves by troubling the clinic staff with their 'silly fears', while a large number of others never went to the clinics at all, either because there was none within reach or because they did not know there was one. They relied for most pregnancy-care information on other pregnant women, and probably got a good deal of grandmother's folklore as a result.

* *International Medical Tribune*, 28 April 1966.

Yet if a miracle happened overnight and all these women did suddenly go for ante-natal care the services would be absolutely swamped.

So there is the second approach. If we are not big enough to shift the boulder we ought to chip away at its weakest cracks. As Allan Barnes has put it:*

> The seeking of pre-natal care is a sociological phenomenon in this country, carrying overtones of fadism in the economically privileged. Providing all women with obligatory pre-natal care will not of itself eliminate mental retardation. Over-concern with pre-natal care as a general therapeutic entity diverts attention from specific factors which have a high degree of correlation with congenital lesions: e.g., diagnostic radiation.

Though he is writing of the U.S.A., the message probably applies everywhere.

What it says, in effect, is go for *specific* diagnosis and treatments – and research. With those poor women, make the most of the fact that they are more likely to have a deficiency of iron, folic acid or specific proteins, and test everyone for these. It says that anencephaly must have a cause: search for it, laboriously, and you may learn how to prevent it more quickly than by trying to give everyone the living standards and health care of Social Class I. Or take all those teratogens. Some scientists have urged that discovering which chemicals do what to the foetus is such a slow, effortful business that the best answer is to persuade pregnant women to put themselves in a kind of therapeutic purdah.

> If a drug affects an adult, it is likely to affect a more sensitive structure such as a fetus. If it does not affect an adult it should not be taken. The incidence of early pregnancy is so high (about 5 per cent of women in their twenties are in their first trimester) ... that the fetus can only be protected by the development of a climate of opinion which regards the casual taking of drugs as a form of sick behaviour which is inappropriate in healthy societies†

* Allan C. Barnes, 'Prevention of Congenital Abnormalities from the Point of View of the Obstetrician', *Proceedings of the 2nd International Conference on Congenital Malformations*, p. 377.

† John H. Edwards, Consultant Human Geneticist, the Children's

So, ideally, pregnant women should manage without tranquillizers, sleeping pills, stimulants, pain-killers, all medicines in fact, should avoid foods containing preservatives or artificial colouring, and should not go anywhere where they might catch influenza, mumps, German measles, measles, or a cold. An impossible way of life, one feels, but all these are suspect teratogens – some on the slimmest evidence of animal experiments, one must add – and they are only a fraction of all suspects. There are very few *known* human teratogens but an infinite array of possibles, and until one clears the innocents the advice is – don't. In its way, of course, it is good advice. But who can possibly live by it? And if no one can, one comes back to the necessity of focusing on the strong suspects and taking notice of *those* – which means a great deal more research.

Of course, both strategies will be used. No politican can afford to ignore the vote-catching strategy of providing good health services or the evidence that wherever they are provided and used infant mortality at least is lowest (as for example in Sweden, Denmark and the Netherlands). On the other hand, no government can afford to ignore research or the pressure to apply it when it discovers a particular way of prevention, like mass vaccination for German measles. But the point that many experts make is that everyone has to realize that the two strategies are mutually exclusive as long as resources and manpower are limited. To some extent we have to choose between them.

Two decades ago the biologist J. B. S. Haldane wrote in a preface to a book on mental defects:* 'At present we can neither stop defective children from being born nor save in a few cases cure their abnormality. We do not know how to do so. Even if we did, we might prefer to spend the necessary effort on the construction of greyhound tracks, jet-propelled bombers, or nylon hosiery.' Fifteen years later he wrote another preface for a

Hospital, Birmingham, England, in 'The Application of Knowledge', one of the series of articles on birth defects published by the U.S. National Foundation – March of Dimes.

*Lionel Penrose, *Biology of Mental Defect* (Sidgwick and Jackson, London, 1948). The second preface is to the 3rd (1963) edition.

new edition of the same book. Now a gleam of hope had crept in. We had made progress, but 'when all the measures of which we can think at present have been taken we shall not probably have reduced the frequency of mental defect by more than about one quarter'. For a very long time, perhaps for ever, we shall have to cope with the stragglers who come out of the womb – the 'valley of the shadow of birth' as someone has called it. Biomedicine has some remarkable tricks up its sleeve that may help us here, but it is also raising sharp and awesome problems. These are what the next chapter is about.

# 7. BIRTH DEFECTS

The control of infectious disease has been one of the greatest human triumphs. Largely because of it, in the developed countries we now expect to live a long life and we no longer expect to bury our children. In the last hundred years in Britain the chance that a baby will die before its first birthday has been cut by six times, that a toddler will die before fourteen by thirty times, and that an adolescent will be dead before fifty-five by five times. We now survive.

If this has been a resounding victory on the whole, it has also raised awesome problems. Without adequate checks to balance it, triumphant death-control has led to population explosions or, where these have been damped down, to old-people booms. But it has also sparked off a third, individually far more tragic kind of explosion: a steady, surging rise in the number of physically and mentally crippled children. This trend is quite new and we are going to find it very hard to live with and very hard to curb.

The trouble is that while antibiotics and so on have hit at several kinds of defect that people *acquire*, like poliomyelitis and meningitis, they seem to be saving rather more people who are *born with* severe defects. Wrapped in cocoons of antibiotics, these usually delicate, susceptible victims of genetic and other congenital accidents now survive. At the same time, doctors and surgeons, committed to the idea of doing all to save and salvage life, are exaggerating the problem by going on where antibiotics leave off. For instance, the one in three mongols with heart defects who now survive once-killing diseases also survive to have their hearts – but not their heads – cured by the surgeon's knife. And so (though many surgeons will not operate) mongols are increasing, and rapidly. In the last thirty years in Britain the number of ten-year-old mongol survivors has more than quadrupled.

It is the same kind of story with many other kinds of defect. The net result of medical progress is an increase in the burden of treating, training and caring for children with long-lasting and often dreadful handicaps that they, their families and society will have to live with for years. And I mean live with, for another thing that progress is bringing is a firm conviction that wherever possible these unfortunates should not be left to rot in institutions but should be reared in the community and at home.

Only a fanatical optimist can believe that this problem is not going to get a great deal worse before it gets any better. In the long run all that we have looked at so far – from genetic warning systems to genetic surgery, from foetal detection for abortion to the growth of foetal surgery – may prevent many birth defects. But it is not going to eliminate them this side of Utopia. And while medical and surgical 'cure power' is growing rapidly it is going to be a very long time, if it ever happens, before we become such wizards at treating or curing all physical or mental defects that they do not matter – that they become as trivial, say, as wearing glasses if one is born short-sighted.

Meanwhile, our fine-sounding attitude that we should do all we can to patch up defects and fit their victims for a place in the world begins to smack of apathy and hypocrisy as soon as one looks at how we actually practise what we preach. Though a lot of people are trying to make things better, they and the rest of us have a very long way to go before we can, in all honesty, face the average crippled survivor without shame. And it may be that we simply cannot afford to go that far. We may have to take other, obvious but perhaps less shameful steps to curb the explosion.

To illustrate this I have chosen a few typical kinds of defect which seem to bring out the main aspects of the problem.

## Surgical cures – pyloric stenosis

Along with many heart defects, pyloric stenosis is the subject of a triumphant surgical success-story, in which a single operation can make a baby with a lethal defect perfectly normal. Pyloric stenosis is due to a genetic fault: the muscle that closes the hole joining the stomach to the small intestine is too powerful, so that when it contracts the victim vomits – often right across the room. It occurs in about 5 of every 1,000 boy babies but only in 1 of every 1,000 girls, and if untreated it kills. The baby starves to death. Forty years ago an operation was invented to snip and weaken the muscle. This effects a total, permanent cure. It is now so routine that it has become the commonest single reason for surgery on infants.

There are no disturbing problems here, except perhaps this: by making a once lethal *genetic* fault trivial, modern surgery is making the 'bad' gene more common. The survivors of surgery now live to reproduce, and from a study by Dr C. O. Carter in Britain it appears that babies born with the defect are increasing at 12 per cent compound each generation.* For us this is no reason for not operating or even trying to prevent the saved survivors from reproducing. But conceivably it might be for our distant descendants.

## Prematurity

Prematurity is the major cause of death in infancy. So, because doctors and their customers do not like babies to die, prematurity is the focus of a massive and increasingly elaborate rescue campaign. More and more premature babies are now reared in the kind of space-age hospital room where you cannot see the crib for the electronic gadgetry around it, or the baby for the electrodes taped on to it and the catheters and tubes that go into it – up the nose for oxygen, into the stomach for food, through the umbilicus for monitoring the blood or giving drugs and transfusions. As one expert has said, X-rays of distressed premature babies are beginning to look like wiring circuits for hi-fi sets.

*In *British Medical Bulletin*, 17 (1961), pp. 251–3. Carter does not give the 12 per cent figure, but it comes straight from his study of how many children with and without the defect are born to people whose lives were saved by the operation in infancy.

This aggressive approach does save thousands of lives and saves them well. Many premature babies just need good nursing in a good incubator to give them a very high chance of living with completely normal lives. But overall, saving them is increasing the amount of residual defect. A recent large American survey* of 20,000 pregnancies has found, for instance, that all premature babies taken together are nearly three times as likely as others to show neurological abnormalities within the first year of life but over six times as likely if they weigh less than 2,000 gm (about 4½lbs) at birth. As doctors get more and more powerful weapons for saving premature babies, these figures are almost bound to get worse because the weapons will be used to dip further and further into the morass of severe prematurity, with all its associated defects like cerebral palsy, epilepsy and physical abnormalities. More babies who would have died without super-intensive salvage and care will survive with their defects uncured or at best patched up.

Most prematurity doctors are acutely aware of this dilemma, and some act to avoid it. Super-intensive care can cost as much as two thousand or even three thousand pounds for a baby that has little chance of surviving, let alone without handicap, while there are never enough machines to go round. So incubators are not all that infrequently switched off to make room for more hopeful cases – sometimes without the parents' knowledge. Nevertheless, by and large what the doctors want is more money for more machines, nurses and so on, so that they are not forced to switch any babies off.

There are some dreadful illogicalities here. For one thing, super-incubators are 'saving' babies that could under more liberal abortion laws be aborted because they had not reached the seven-month gestation point where, tradition says, they first be-

* Preliminary results (reported April 1968) of the Collaborative Project on Cerebral Palsy, Mental Retardation, and Other Neurological and Sensory Disorders of Infancy and Childhood, run by the U.S. National Institutes of Health. The actual incidences of abnormality found were 2 per cent for all 20,000 pregnancies; 5·7 per cent for all premature babies (under 2,500 gm or 5½ lbs); and 12·5 per cent for babies under 2,000 gm at birth.

come viable if independent of their mother and abortion becomes manslaughter or murder. I have seen in an American hospital a distressed 'baby', on which about a hundred dollars a day was being spent, which was four months premature and weighed only one and a half pounds. No one thought it had much chance of pulling through, and the parents, frightened of the bill and residual defect if it did survive, wanted the machine switched off. If the 'baby' had been inside the mother she could have had an abortion; now it was born, the doctors took the save-at-all-costs line.

The greatest illogicality is that this heroic save-all approach is almost bound to produce *more* handicapped children. When a premature baby is born the doctors cannot diagnose possible brain damage, so it goes straight on the machines to give it a chance. Only later, if it has not died, do mental defects appear. But there is now some fairly strong evidence* that premature delivery is more often the *result* of abnormalities in the foetus than the cause of later handicaps. It is not so much a too-early birth and a failure to provide the best post-birth environment that causes epilepsy, cerebral palsy and many physical abnormalities of the 'prematurity syndrome'; these defects very often *cause* prematurity (perhaps through a hormonal disturbance in the mother caused by the defect). So however good the super-incubators, using them can only preserve defects – unless doctors *and* parents are more willing to cut their losses and switch them off.

## Chemical 'cures' – phenylketonuria

Many defects can be treated by continual dosing of the victim's biochemistry – with drugs, hormones, special diets and the like. The diabetic's insulin shot is the classic example; thyroid hormones to stop the child who cannot produce thyroxine from becoming a goitrous cretin is another; blood transfusions or injections of clotting agents for the haemophiliac is a third. There

* C. M. Drillion to the 1st Congress of the International Association for the Scientific Study of Mental Deficiency, Montpelier, September 1967; reported in *World Medicine*, 31 October 1967.

is a spectrum here from the relatively easy, cheap and wholly successful cure of the first to the difficult, expensive and far from successful 'cures' of the other two. Haemophiliacs can now survive well, but they and their families have a hard time of it. And thyroid hormones far from always prevent cretinism.

Specific chemical defects are probably the cause of most severe mental subnormality, and there is now a widespread belief that biomedicine will soon dramatically improve our chemical cure power. We will come to that in a moment. In the meantime the point is that many experts are saying that our chemical cure power is already so effective in some cases that alternatives to using it – alternatives ranging from genetic prevention to abortion and infanticide – should be unthinkable. Phenylketonuria (P.K.U.) is the example they usually cite.

This recessive gene defect occurs in about one or two in every 10,000 babies (or 100 to 200 times a year in Britain). The victims lack a single enzyme called phenylalinine hydroxylase that is normally made in the liver and converts phenylalinine – a constituent of most protein food – into a chemical called tyrosine. If this is not treated, from birth onwards phenylalinine and other toxic products build up in the bloodstream – some to be excreted in the urine but some to 'poison' the brain. At about six months the baby becomes noticeably retarded, and if there is still no treatment the vast majority will become idiots (I.Q. under 25) or imbeciles (I.Q. 25 to 50). A lucky 2 per cent may have a near-normal I.Q., possibly because they have a harmless variant of the defect.

Most developed countries now have mass-screening programmes to detect P.K.U. babies. In some states of the U.S.A. they are compulsory by law; in Britain, where they are merely recommended, they have nevertheless become more or less universal. The usual detection method is the Phenistix test. When the baby is about six weeks, a strip of paper coated with a chemical is pressed into a wet nappy and if it turns greenish-blue the baby has probably got the disease. Sadly, it is now known that this quick, cheap test has been missing between one and two cases for each one it picks up, probably because at six weeks the 'poisons' are too dilute for certain detection. Since one cannot delay the test much longer in case damage has already been done, a surer, early test is badly needed. The answer is the Guthrie test, which utilizes the fact that phenylalinine in a victim's blood slows the growth of a bug called *Bacterium subtilis*. This test involves taking a drop of blood from all babies by a heel stab, drying it on filter paper and sending it to a central laboratory. It can be made as early as ten days, and from wide use in the U.S.A. is known to be totally reliable. The administrative difficulties and costs of the Guthrie test, plus the problem of whether the public will accept heel stabs for their babies, are now severely exercising many health authorities, including the British.

Once detected, a P.K.U. baby has to go on a special diet low in phenylalinine. If this is started by about three months *and adhered to rigorously* there should be only slight or no loss of mental function. But since phenylalinine is in so many foods, keeping to the diet rigorously is far, far easier said than done. There are now special synthetic P.K.U. diets which help, but to keep a child on them can be a formidable job – not at first, of course, when the doctors are saying that given the diet the baby will be all right and the baby is only worried about getting something in its bottle, but later, when the baby becomes a child and the parents are totally committed to the treatment.

At best, as a recent American study has shown,* the special diet can produce severe emotional, psychological and social stresses for all concerned. A child who is banned from eating milk, ice cream, meat or fish, cakes, eggs, bread, vegetables and even chocolate – except perhaps in small, carefully weighed amounts – faces fearful strains. Many cannot go to friends for tea in case temptation is too great; others try to trade toys for food with their siblings and friends; stealing food from the refrigerator is a constant parental complaint; school meals are a dreadful problem. At home and school, the child is constantly reminded of his difference and the fact that he is walking over a chasm of idiocy on a dietary tightrope. Parents, on the other hand, are strained either into being very strict – no eating out, no friends for tea, and so on

* A. C. Wood, *et al.*, 'Psychosocial factors in phenylketonuria', *American Journal of Orthopsychiatry*, 36 (1967), pp. 671–9.

– or their guilt and pity lead to an occasional relaxation. How *could* milk and eggs be harmful just this once?

At worst the strain can lead to stopping treatment altogether. Quite a few P.K.U. children have lapsed into idiocy because their parents could not keep it up, while paediatricians have even had to advise them to give up the diet – knowing the outcome – because they knew the parents were incapable of controlling the child's lust to eat normally. The moral is that biologists and doctors who talk about 'perfect cures' for biochemical diseases should remember that children are more than mere biochemical machines, and their parents more than machine-minders – a moral which will occur again in various guises in the rest of the book.

It would not be so bad if the diet could be stopped after a time: a stressful childhood might be a fair price to pay for a normal adulthood. But at present no one knows whether it can be. Some biologists suspect that it could be stopped since the body may develop tolerance to high phenylalinine levels. But how does one test this when the test may turn a 'normal' child into an idiot? Though the experiment has been made for a short time,* no one has yet dared take a P.K.U. victim off his diet for more than a day. So the terrible restriction of the 'cure' at the moment has to be a lifetime burden, and the idea of starting it in the first place becomes that much less defensible.

### Treatment 'cures' – cystic fibrosis

Cystic fibrosis is the prime example of a congenital defect in which handicapped lives are being prolonged by more and more elaborate and *continuous* treatment. Because, once started, it has to go on for years, and the cost and stress of treatment, as in the case of kidney machines, accumulates dramatically each year.

---

* In 1966 at the Hospital for Sick Children, London, the diet of twelve P.K.U. children was stopped temporarily so that they got a short, high-level burst of phenylalinine in their system. Seven of the twelve produced abnormal electroencephalogram readings soon after, while the other five showed only mild changes. The researchers concluded that the special diets should continue in all cases.

The disease is the commonest of all serious recessive gene defects and affects about one in 2,500 babies. Basically it is a disorder of the pancreas which leads to abnormally viscous mucus in the pancreas, lungs and sometimes the intestines. It also causes very salty sweat and saliva and, because of this, simple tests are now possible which can detect all sufferers soon after birth and may lead to mass screening of all new-borns.

FIGURE 7.1   Cumulative survival curves for cystic fibrosis

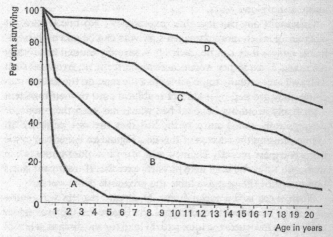

Adapted from *World Medicine* (16 May 1967).

Untreated, the average life expectancy used to be seven months. As Figure 7.1 shows (Curve A) over 70 per cent died in the first year, more than 95 per cent were dead by five and survival after ten was extremely rare. The lungs, progressively filling with fluid, are either destroyed or in milder cases are open to infection by pneumonia. With continual antibiotics for infection, and enzyme treatment for the pancreas defect, the survival rate is much better (Curve B). In the last few years a more rigorous treatment regime has been introduced for severe cases. To help to keep the lungs clear they are periodically 'washed out' by fine mists, either from aerosols given by face masks or – more usually – by putting the baby or child to sleep every night in a mist tent. Some children have been on this regime, in hospital or at home, for several years. With very severe cases the washing is done by putting a tube into the lungs under anaesthesia. With this new treatment survival is more frequent (Curve C). If it is started very soon after birth, when the new sweat test can now detect the defect before there is any appreciable lung damage, survival is still further improved (Curve D): the average life expectancy has shot up to about twenty-one years.

Or should one call it a 'life' expectancy? No one is sure yet whether or when more damaged survivors can be taken off treatment. Unless they can be, their life is severely though not totally restricted. Even if they were otherwise perfectly fit, freedoms such as travel are virtually impossible: like the man on the kidney machine, they are tied by a kind of umbilical cord to their mist tent – probably around a hospital bed which has been their main or perhaps only home since birth. But they are not perfectly fit. While praising the success of this new regime an American cystic fibrosis expert recently claimed that *often* his older patients can work full time and *even* take physical exercise. If treatment starts late or is not the best available the prognosis is far worse.

Then there is the social stress. Numerate readers will realize that since treatment is continuous, in Figure 7.1 the area under each curve measures the total patient load for the disease, while of course going from Curve A to Curve D also raises the cost per patient dramatically – from near zero for no treatment to a few hundred pounds a year (mainly for drugs) for the most severe cases on full survival treatment. Though these costs are far from astronomical, many paediatricians have grave doubts about the morality of massive efforts to shift all sufferers from Curve A to Curve D regimes.

## Surgical 'cures' – spina bifida

This defect raises some of the gravest practical, social and moral problems in all medicine. It is also a supreme example of our hypocritical failure to face the consequences of the medical ethic that insists that all should be saved.

About three in every thousand babies – or roughly three thousand a year in Britain – are born with a backbone that has not properly folded in to enclose the spinal cord. It is still split, or bifid, so that on the backs of most of these babies there is a patch like an opened kipper with part of the spinal cord lying in it, naked and vulnerable. Faced with such a disaster at birth, there are only three things one can do: kill the baby; do nothing, hoping that it will die; or operate as quickly as possible. None of these possibilities, not even the last, is free from grave dilemmas.

Leaving out killing for the moment, if one does nothing the prognosis is very nasty. Four fifths of the babies will mercifully die within a year, but the rest, the tough ones, will live on. Virtually all will be paralysed from the waist down, and incontinent because of damage to their exposed nerves. Four out of five of these survivors will get hydrocephalus; their heads will swell out, some until they are too heavy to hold up. Severely retarded, often spastic and blind, they will spend their childhood in institutions that most of us do not care to think about, let alone visit. By adolescence virtually all will be dead.

The third alternative is *immediate* surgery. About fifteen years ago reasonably efficient valves were invented to drain off the build-up of cerebrospinal fluid that causes hydrocephalus. Since about 80 per cent of spina bifida babies need them, now that they are available surgeons can work on trying to patch up the spinal damage. But unless this is done within a few hours of birth the damage may be irreparable: the patch on the back will swell, tug at the cord and pull out nerve roots. Meanwhile there is a growing risk of meningitis.

Surgery can be spectacularly successful. Take the case of 'Peter'. He had a four-inch-square swelling on the back of his head – a mild case, but a nasty enough sight for Peter to have been rejected – left to die – until a few years ago. Instead, surgery and devoted nursing helped turn him into an alert, enchanting child with hardly a hint of physical disability. Conversely, the results of not operating early can be heart-breaking. Take the case of 'Sarah', who had been kept at home without treatment. She developed severe hydrocephalus. At fourteen weeks, when she still had not died (as her parents hoped) she was taken to hospital for treatment. It was too late: she died there. But the surgeon who would have operated reports that before she died he saw her legs moving. She was not paralysed. If she had had early surgery she could have developed into a near-perfect child like Peter.

Fired by cases like these, surgeons are being forced into advising and doing early surgery on *every* spina bifida baby. Because at birth they cannot tell which babies will turn out well and which badly (except with a grossly damaged or premature 5 to 10 per cent) they feel they have to operate on all. Yet they also know that doing this, while saving some Peters, will also salvage many more who will be permanently and severely crippled. As one surgeon told me: 'Sometimes I feel we've gone too far. I know that I have kept some alive who would have died otherwise, and would have been better dead.' Another surgeon said, 'Active treatment is the easy way out. Most of us don't have the moral courage to think of other approaches.'

Despite these qualms, the proportion of spina bifida babies going for immediate sugery is rapidly rising. So, inevitably, is the number of long-term handicapped survivors. So is the medical, educational, social and family burden they involve. Figures on the size of the problem vary widely. It would perhaps be generally agreed that of the three thousand spina bifida babies born each year in Britain about one thousand are stillborn. This, at least, is the traditional stillbirth rate for the defect. But many of these deaths may not be natural at all. Obviously there are no infanticide statistics, but there is the curious case of the county of Northamptonshire. Since 1962 virtually all spina bifida babies born there have been taken for immediate operation to a famous London hospital; and also since 1962 the 'stillbirth' rate for spina bifida has been virtually zero. Before 1962 it was the usual one in three. Whether parents or midwives do not now kill those one in three babies because spina bifida is supposed to be treatable and they do not *want* to kill, or whether parents *cannot* kill because the immediate removal of the baby gives them no opportunity to, is not known, The difference is rather important.

Anyway, assuming the one in three stillbirth rate, 2,000 spina bifida babies survive for surgery. If all are operated on by the most skilled teams, about 1,200 (60 per cent) will survive to school age and beyond. Of these about 500 will be mentally normal and have physical defects so slight that they can lead near-normal lives. Another 500 will also be mentally normal but will be to some extent – and mostly to a great extent – paralysed below the waist and incontinent. The last 200 will be severely handicapped physically but will also have brain damage making them subnormal or ineducable. So every year on a conservative estimate surgery is salvaging about 1,200 long-term survivors, of whom 700 must have from society and their families the most intense and specialist care and education.*

The stress of rearing a spina bifida child, even though he is mentally normal, can be dreadful. Back home at five weeks or so, the baby is much like any other: normally wet, dirty, unable to walk, cheerful or tearful. But in the next five years, it has been estimated, the majority of survivors spend on average two years in hospital, mostly in week- to fortnight-long visits. The hydrocephalus drain must be checked, cleared if it blocks, and lengthened as the child grows. New leg braces must be fitted and the child taught to use them. Physiotherapists will try to get the maximum use out of muscles that are working. There may be lengthy tests for possible bladder or kidney damage. Girls may need an operation to divert their urine out to a spout, roughly over the appendix area, where it is collected in a bag. Many spina bifida boys have to wear a plastic bag over their penis. Because specialist centres are not thick on the ground, journeys to the hospital of two hours each way are not unusual – all with a paralysed, incontinent toddler.

* Actually, this is a very conservative estimate. Most spina bifida babies do not get to the most skilled surgical teams, so for the country as a whole the amount of handicap is higher: only 25 per cent with minimum handicaps rather than the 40 per cent given above has been suggested as an overall figure. And if the 'save all' policy of say Northamptonshire was followed everywhere so that all 3,000 went for surgery that 700 would rise to about 1,000.

Poor housing – an upstairs flat, a shared lavatory, lack of running hot water – is hardly compatable with caring for an incontinent child in leg-irons, while local authorities can often be sadistically callous about the needs of parents of such children. In Britain recently a couple with a spina bifida child who had only a garden-shed lavatory asked their local authority for a two-hundred-pound grant towards proper sanitation, but were refused on the grounds that the house was due to be demolished in fifteen years' time. This may be exceptional, but nearly all spina bifida parents – like the parents of all severely handicapped children – report that they feel desperately cut off from the really considerate, expert week-by-week help and advice they need.

At school age new problems begin. Because the 'sooner the better' policy of surgery also applies to training for incontinence and paralysis, the trend is to put spina bifida children into special boarding schools at younger and younger ages. However, in early 1967 there was only one school (with forty-five places) in the whole of Britain especially for incontinent children, mostly spina bifida cases. Two more were opened during the year, bringing the total of places to around 150. These schools are excellent. They are well staffed, and apply an enormous amount of skill and devotion towards minimizing the effects of the children's handicaps. By exhaustive exercise, physiotherapy and bladder-training regimes, children who enter the schools bedridden and written off as totally paralysed leave them walking – sometimes even without irons – and their incontinence either cured or at least not socially embarrassing. No amount of training, though, can do much for the children's future sex life: few will ever achieve an orgasm, and they are unlikely to father a child, though pregnancy is possible.

One of these schools has now opened a nursery wing to take two-year-olds as boarders. Mothers, the argument goes, cannot watch fluid intakes, or infection, or treat the usual crop of nappy sores: medically speaking, the best place for these toddlers is school. This is the inevitable result of the doctor-dominated 'get them early' philosophy that now applies to spina bifida. Psychologically speaking it may be a bad bargain.

Most spina bifida survivors cannot find a place in these special schools and are sent instead to mixed handicap schools. Luckily for the authorities they can take the places prepared for the polio victims who now no longer need them, thanks to mass immunization. But though administratively convenient, this is a cruel policy. Even if the mixed handicap school gives (and often it does not) the special training that all experts say makes the whole difference to the future quality of a spina bifida child's life – his ability to walk and conquer incontinence – few of the other children will share his singular disability, the fact that even in adolescence he is often wet and smelly. In mixed schools spina bifida children are often social outcasts.

Bluntly, in no society are more than a tiny proportion of spina bifida victims getting the care and treatment that can make the best use of their potential. And it does not look as if they will for a long time to come. If Britain were to provide special schools for every spina bifida child its surgeons are now salvaging it would have to build one with fifty places in it – and staff it with about ten skilled people – each and every month for fifteen years.

The most acceptable method of minimizing the dreadful toll of these defects – and the many others not mentioned here – is clearly to find total cures for them. Obviously no one can predict what specific cures there will be, or when they will come, or how effective they will be, and so on. Nevertheless, the biomedical world is excited by the general belief that much will be possible fairly soon – so much so that people are coining grandiose names for the paediatrician of the future. Rather than a child doctor he may become a biological engineer or, if Professor Lederberg has his way, a euphenicist.

## Defect engineering

No one really seems to expect dramatic progress in surgery for structural defects. Operations that can be carried out now will be improved, so that fewer babies die from them and hospital stays are shorter. But by and large, chipping away at today's more drastic, intractable anomalies – like the more complex heart defects –

will be a very slow, arduous job. Transplants, especially of the liver, could save many babies when (or if) they become routinely successful; while it is just possible that neurosurgeons may learn how to graft and join nerves so that they all link up in days or weeks rather than a few of them doing so in months, as now. This could have a profound impact on spina bifida treatment, for instance.

For the physically handicapped there will certainly be astonishing advances in all kinds of implanted or attached electronic-mechanical aids. With spina bifida children, again, there could be transistorized bladder controllers; implanted amplifiers to boost the nerve signals that do travel to their leg muscles, only much too weakly to make them move; or even powered 'exoskeletons' of metal and plastic strapped to the legs and wired into the nervous system so that, say, a twitch of the shoulder sets the left leg walking. But despite the boost the thalidomide affair gave to this sort of technology it is still abysmally under-supported. There *are* new advanced prosthetic devices, and some are even on the defective limbs of children, but as long as every society is a million times keener on packing its electronic skills into weapons, cars and machine tools, every newspaper picture of a new, wonder, prototype prosthetic device is an almost obscene joke unless the caption announces a massive financial commitment to it. However, it seems clear that for the foreseeable future artificial aids will be troublesome, noticeable, poor substitutes for the real thing.

What the promise of euphenics and defect engineering is really about is *biochemical* medicine – a vast extension in the quantity, quality and extraordinariness of the kinds of treatments that we use to help the diabetic, the haemophiliac, the phenylketonuric, or even the slum child who might get rickets but does not because he swallows a spoonful of cod-liver oil. In other words, therapies which we now take for granted will be stretched to the limit by the new insights and skills of the molecular biologist.

Broadly, there are three main approaches to the future cures of biochemical defects. The first we have already met with the phenyl-ketonuric. Because an enzyme is missing, the substance on which it normally works accumulates and does damage, so the answer is to cut out that substance. This will certainly be done with many more biochemical diseases where no one yet knows what that substance is. But this is hardly a satisfactory cure-all solution unless the restriction is trivially unimportant.

In other defects the fault is that too little of some vital chemical is produced. Examples are goitrous cretinism, haemophilia and diabetes. One approach here is simply to supply what is missing. Unfortunately, even if one knows precisely what to supply, this cannot always be done because supplies are short or, in some cases, the body will not absorb the chemical easily. The breakthrough for diabetics was not so much the discovery that they needed insulin but the realization that insulin from horses would work. Similarly with haemophilia: the old days when a bleeding haemophiliac had a queue of blood donors outside his hospital ward have been transformed by the discovery that the clotting factor he lacks and that he was supposed to get from all those donors can be concentrated by rapidly freezing blood. Now all the haemophiliac needs is a shot of anti-haemophilic globulin.

There will be a lot of progress here, some of it bizarre. Where animal sources will do there should be few supply problems. In some cases, though, there is bound to be difficulty over species specificity: animals will make the chemicals, mostly proteins and enzymes, in slightly different forms from the human kind. This could give trouble with long-term treatment, so that the chemicals might have to be extracted from human organs after death. Artificial synthesis would be a more acceptable alternative, and it is likely that this will come on a fairly big scale. The difficulties are formidable. It took six years for Frederick Sanger to produce the first chemical formula of a protein – insulin – and another eleven before insulin was first synthesized in 1966 by the Chinese. And insulin is a very simple protein, with only 777 atoms in it. Nowadays, with automatic tools for discovering the structures of these enormously complex substances and others for putting the right bits together in the right order for synthesis, the job is a lot easier. But even the optimists think it will be a long time before the simplest ones can be made at all cheaply, or the complex ones like haemoglobin made at all. Another possibility might be to harvest human enzymes in big vats of cultured human tissues, with, say, a pancreas vat here churning out insulin, or a spleen vat there making anti-haemophilic factor for 'bleeders'. This, too, is going to be far easier said than done. Though growing cultures of human cells is now a biological commonplace, almost without exception what is being grown is a bunch of cancer cells. To keep a culture of specialized cells from going cancerous may be exceptionally difficult. So might getting them to start growing in the first place: for example, liver, pancreas and central nervous system cells – which make many crucial genetic disease enzymes – multiply at a very low rate in the body, if at all, and very quickly stop growing in cultures. Some biologists guess that the only solution might be to prime the vats with organs from human embryos, whose cells do proliferate rapidly.

Apart from supplying missing products, a second technique for 'curing' these product-deficient defects could be to switch genes on or off. One of the fundamental biological discoveries of the last few years is that, though every cell is equipped with the same set of gene instructions, what makes the chemical activity of this cell different from that one is the way complicated feedback loops in the cells stop some genes from 'working' and stimulate others. There is much talk of the idea that it will one day be possible to modify this gene 'induction-repression' or 'switch on, switch off' system to cure many defects. For instance, in a defect where the liver cells fail to make an essential enzyme it may not be because the structural or 'know-how' genes are missing, but because the switch-on circuit is defunct. In which case it may be possible to switch it on either in the liver or even in another more accessible group of cells. Indeed, there are one or two rare genetic diseases where this kind of thing has been tried on humans, by drug treatment, and with some success. About all one can say now is that more successes are expected – some time.

The third main bio-engineering approach is the most radical of all. So far every kind of treatment I have mentioned has meant or will mean regular, day-by-day dieting or dosing, with all the life-limitation that usually involves. Ideally, the perfect 'cure' would be permanently to change the victim's chemistry at its most basic level – where the root defect is. For example, with the phenyl-ketonuric, one provides a permanent supply of the enzyme he lacks, phenylalinine hydroxylase. Because enzymes are catalysts, or chemicals that speed up reactions without themselves being broken down, there is no need to give them in vast amounts: on the contrary, a pinch in the right place could in theory last for decades. On the other hand, because enzymes work best (or only work at all) inside the correct organs, surgery will be necessary. There is no question of swallowing an enzyme elixir or rubbing in a permanent embrocation.

Several techniques may be possible. One is to wrap up the enzymes in plastic micro-capsules so that they can slowly seep out into the bloodstream. This is not as good as putting them in the correct organs, but from dog experiments it seems to work tolerably well (at least with one enzyme that has been tried). The problem is to control the seepage rate, so that the enzyme levels in the body are correct and steady. Capsule technology may solve this in time, but at present the only feasible method seems to be to have the capsule holder outside the body and to pass a trickle of blood through it along plastic tubes joined up to an artery or vein. Another possible technique is the transplant. Phenylketonurics, for example, should theoretically be cured by a liver transplant, because it is in the normal liver that phenylalinine hydroxylase is made. Similarly, the haemophiliac might be cured by a spleen transplant, and there is a whole list of genetic diseases which could be cured by transplanting a kidney. Further off there is the possibility of taking the relevant tissue from a patient, doing some genetic engineering on its cells, and then re-implanting the tissue where it came from. In this way a part of the body would be re-programmed into normal genetic and therefore chemical functioning.

All this gives a hint of our future power to cure defects. But until each possibility becomes a practicality no one can predict how successful, how affordable, or how acceptable each technique will be. In the meantime, caught between the past, when nature used to solve the problem by death, and the future, when the doctor might in more and more cases offer acceptable cures, how do we cope with the growing burden of defect?

At present every part of official public morality – the law, medical codes, the Church, etc. – supports absolutely the right of every baby to live. Officially we all totally reject the Spartan solution, which was to expose the deformed and monstrous, as too barbarous. Officially we totally support the Christian ethic, which holds that at least after birth all life is sacred and even the near-vegetable Cyclops-child with its one eye in its forehead has an inviolable right to live. Though society can kill when it is threatened, the micro-societies or families of which it is composed cannot – even though they feel deeply threatened by the arrival in their midst of a monstrous invader.

In private, of course, we think and act very differently – and with a vast range of differences. Some people claim that they gain a profound spiritual uplift from rearing a severely subnormal child, whatever the cost to themselves or their other children, and whatever the outcome for the child itself. Others – I believe often with less false sentiment and more genuine humanity – think it better to avoid the agony (not least for the child) of trying to turn even a moderately severely defective baby into a 'happy and useful member of society', to face the agony of killing it while newborn, and to get on with having another normal baby, a more rewarding focus for their love and devotion. Most people, rather uneasily perhaps, fall somewhere in between. Because of this range, the law and other guardians of official morality turn a blind eye to infanticide, unless someone is too indiscreet, or some widespread disaster like thalidomide brings its likelihood to public attention.

I do not intend to argue all the pros and cons of infanticide and its only alternative, the fully-fledged 'save and care' ethic. Logically, all the arguments are for infanticide, and for diverting the gross exertions we make to keep alive children who will never be normal towards doing more for normal children. While we save the abnormal, normal children starve for food, love and education, while some, in the ghettoes of our most 'civilized' cities, are eaten alive by rats. Emotionally, most of the arguments are against infanticide. We still pay at least lip-service to the idea that all human life is sacred, and our guilt and fear about killing dominates other considerations. There are also entirely untested fears that to allow infanticide would erode belief in the sacredness of life and disturb the stability of society.

What I do suggest is that biomedical progress *forces* us to choose one alternative or the other, in each case, much more openly and less hypocritically than we do now, and that having chosen we must act honestly and with a full commitment.

Take infanticide first. I do not mean by the word infanticide anything involving selection by the state of who should live and who die (who would choose, where would we draw the line?), though our descendants may come to this in time and by common agreement. Nor do I mean the mercy-killing of children. By infanticide I mean the killing of a baby because its parents make a hard-thought-out decision that to keep it alive involves a greater sacrifice for all, and is more offensive, than killing it. Almost invariably this will mean killing a newborn or very young baby – killing, in other words, before the total commitment point where the parent-child bonds get too strong and the parents decide (for better *or* worse) that they must make the best of it. Very roughly this applies to the half of all known defects – including mongolism and spina bifida – that are recognizable at or very soon after birth, the other half appearing only at later ages. However, medical progress, with its drive for early diagnosis, will increase the number of candidates. When the P.K.U. child is tested at six weeks he may be smiling; when he has a Guthrie test in the first week he is not.

Perhaps we are going to have to make a chronologically minute, but emotionally vast, shift in the *dating* of our acceptance of a baby into society. Traditionally we do this at birth, as the hidden foetus becomes a visible child. It is from this point that a human being acquires his full human rights. Perhaps this traditional moment is not the most appropriate one. It may be that two or three days' or weeks' 'probationary' life should be accepted as a period during which doctors could check for defects and parents could decide whether or not they wanted to keep and rear a damaged baby.

Nowadays, if parents wish to kill a defective newborn, there is little they can do about it. Most babies are born in hospitals. If they are born at home, damaged ones (as those cases of spina bifida emphasize) very soon find themselves in hospital – and under the umbrella of the hospital ethic. Even their second way out of the problem – abandoning the child to state care in an institution – is coming to be more and more unacceptable except in the worst cases. It never was much of an answer, since it replaced the short-lived guilt of infanticide by the life-time guilt that the child would have been better dead or should have been kept at home. Nowadays places in good public institutions rather than gruesome, understaffed, overcrowded warehouses for the subnormal are increasingly hard to find, while private care is expensive. Furthermore, there is a strong trend to advise against institutional care anyway, except for really bad cases or where the family clearly cannot cope. The old advice of the doctor – 'Don't waste your time, put him in a home and have another' – is being steadily replaced by, 'Keep him at home, put all you've got into him, and you'll be surprised at how much you can do for him.'

So while medicine is pressing parents to keep their handicapped children, the opportunities for parents to choose not to keep them are being reduced. Infanticide is becoming more and more the doctor's choice. Or, rather, the decisions about it are increasingly made under the ethical umbrella of the hospital environment, where for an individual doctor to go against the common code is exceptionally difficult, psychologically and practically.

In public the medical profession puts up a united front in its dedication to preserve life at all costs, but in private the façade often cracks. At a recent open conference on the cost of life* a group of doctors agreed yet again that the duty of the medical profession was 'to preserve life and promote the welfare of their patients whether the latter were malformed, mentally handicapped, tetraplegic, renally bankrupt, or just elderly'. Afterwards most of these doctors to whom I talked privately said that with

* *Proceedings of the Royal Society of Medicine*, 60 (1967), pp. 1195–1246.

infants it was often better to 'knock them off' and that 'duty to patients' is an empty phrase when one is considering not so much a handicapped child as a handicapped family. Doctors are far more ready to talk of mongols as hopeless idiots and wards for the subnormal as 'vegetable gardens' than most people suppose.

The most responsible doctors steer a middle course, acting as adviser and agent for the parents. They realize that whether to prolong or end the life of a severely defective child is far more than a medical question and that the primary right of decision is the parents'. What they then do about it, though, is quite a different matter. Few doctors will kill, though they may not 'strive officiously to keep alive'. The result of this uneasy conflict between killing and letting die can have horrifying consequences. I have talked to several surgeons who admit that they ought to have killed some babies in their care, because they and the parents felt it right, but could not. There was no chance of withholding treatment either. So a baby nobody wanted lived. Others have admitted to waiting for an infection so that they could withhold antibiotics – but the infection never came. Worse still perhaps are the even more devious ways of evading the quickest, humanest death – such as a lethal injection – because the doctor lacks the courage of his convictions or has to consider what the coroner might think. Surgeons have told me that they have put spina bifida babies through severe, high-risk operations hoping they would be lethal. One said he once switched off an incubator hoping to chill a spina bifida baby to death, but when it had not died after a night he turned the heat full on. That worked, after a time.

Then there are the doctors who do *not* take into account, or ignore, the parents' wishes. Some claim they always try to have the time, opportunity and patience to talk at length to parents, and the humanity and courage to go against their own instincts if the parents ask them to – whether this means saving or killing the baby. More usually there is no consultation (or a minimal one) and no question of parental consent. One spina bifida surgeon has written recently, without any qualification at all, that 'one now sets out to save all'. Another, who works in a poor, industrial area of Northern England, is said to have an exceptionally high post-operative mortality rate that is nothing to do with his surgical skill. He assumes that the families of his patients cannot and do not want to cope, and that is that. Both these attitudes are ethically unacceptable but the latter – in a situation where society ensures that parents will feel dreadfully guilty if they even suggest that the child should not be saved – may be the kinder.

I sense a swing in attitudes among doctors – a slow, hardly definable shifting of the individual props that support the 'save-life' ethic. Many doctors in the front line of the salvage campaign are becoming alarmed at what they are doing. Partly it is a matter of growing numbers swamping resources. As one mongol expert told me, 'My wards are getting so full I can't even keep a good vegetable garden.'

Mostly, though, it seems to be due to a slowly spreading realization that salvage can often be an expensive and futile road to misery. A leading London children's hospital surgeon told me that he and his colleagues are terrified in case some well-meaning person gives them a kidney machine publicly, so that they can hardly refuse it, and they have to use it on patients better left to die. Another dreads a lobby to provide more resources for more cystic fibrosis sufferers to get full treatment. There is no way of making a meaningful count, but medical conferences and articles seem to be producing rather more often recently statements like this:*

What can be an unhappier result than that of a persistent faecal incontinence found in a 12-year-old girl, whose life at birth was victoriously saved from the results of an imperforate anus, but whose anal sphincter, if it ever existed, was lost in the life-saving operation ... The handicapped child, life successfully saved, may well hate us in the future as it faces the competitions of life when the adolescent and the adult pressures begin to weigh in upon it. I am reminded of the problems we faced in acute poliomyelitis when, with the respirator, we began to save triumphantly the lives of those who had respiratory muscle paralysis, only to see some of the most tragic situations imaginable, grossly handicapped, yet living individuals, utterly dependent on others constantly for their existence.

* Dr James L. Wilson, 'Congenital anomalies in modern clinical medicine', 1st International Congress of Congenital Malformations.

Meanwhile the Christian ethic is paramount, and to advocate the vóluntary mercy-killing of babies is like campaigning for state Belsens. And so all states, most people and most doctors go for the second alternative – save handicapped babies and care for them as best one can. And how well do we do this? Occasionally exceptionally well, on the whole not at all well, and quite often abominably badly. There is no point here in giving lurid descriptions of hospitals for subnormal children where incontinent idiots literally wallow in their own excrement because there are too few nurses to care for them properly. Nor need one describe the few exceptional places where loving, devoted staff and specialists put hours of effort into getting a single child to take another small step out of backwardness – perhaps to walk up a stair, or say a word, or write a single letter. There is no point because these are the extremes.

Instead let us take as an example a brief, general look at the provision made in Britain for mentally subnormal children – one of the largest overall categories of handicap.

As we have seen, there is a widespread campaign to persuade the parents to keep such children at home. There is evidence that given individual, warm attention, a far greater level of performance can be expected than from a child with a similar handicap reared in an institution. So it is best for the child to be kept at home, and it is left to the parents to try to carry the burden of caring for the child without totally ruining their own lives or the upbringing of their other offspring.

This is a formidable task, but does the state put much weight behind the campaign? In a survey* of 87 London households containing a subnormal person, 29 depended on old-age pensions or National Assistance, another 23 with wage-earners found it very hard to manage on their incomes, and with 34 families the house was unsuitable – in one case a chair-bound thirty-five-year-old lived in a basement with a bathroom two flights above her, in

* Jean Moncrieff, *Mental Subnormality in London* (P.E.P., London, 1966). See also *Stress in Families with a Mentally Handicapped Child*, the report of a working party by the National Society for Mentally Handicapped Children, 1967.

another an incontinent fourteen-year-old mongol lived in a house with no bathroom and an outside lavatory. Though the families on the whole accepted the need to look after the subnormal person, many said they were very isolated – no one would baby-sit with their 'child' – and the subnormals themselves were more or less imprisoned in the home. About half went to training centres or sheltered jobs outside the house, but all spent their leisure time strictly at home, mostly watching television and playing children's games. Too often, the families reported, case workers who tried to help them were impersonal and underestimated their difficulties. Though this could all be changed, until it is (as the *Lancet* has said) community care will remain a euphemism which, for the family, is no joke. Commenting on this survey, Jack Tizard, Professor of Child Development, University of London, has said that if anything the condition of handicapped families looks as if it is worse now than it was seven years ago, when he carried out a similar survey.

Some children, at first kept at home, will eventually be placed in institutions. Others will enter them very early in life. What provisions are being made for them?

The first point to be made is that there are not enough beds provided in *any* kind of institution, let alone the best, to give parents a real choice: some severely subnormal children *must* be kept at home whether their families accept it or not.

In 1971 there were about 64,000 beds for mentally subnormal patients of all ages in British hospitals and they had increased by something like 10 per cent in the previous decade. This rise was partly due to greater provision to shorten the waiting lists, but also to an absolute rise in numbers, mainly of children and adolescents. Yet while many of these hospitals are grossly overcrowded, more than 60 per cent of the patients in them need not be there (having no other handicaps) but are forced to stay because no other residential accomodation is available. Meanwhile an estimated 97,000 subnormals are known to live at home, with perhaps another 100,000 who are unknown to the authorities (according to a 1971 estimate by the CARE organisation). A majority of the relatives of these subnormals put up with enormous strains and self-sacrifice trying to cope with them at home – the incidence of breakdowns among relatives is high – and would do anything to find a hostel or some other residential accommodation for them.

Within most of these institutions there is a grave shortage of doctors, nursing staff and teachers. Few trained staff choose to work in surroundings which are often depressing, and where, because there are too few of them, they can do almost nothing for their patients. As a result there is much less flow *through* hospitals for the subnormal than there should be. Adolescents who, given intensive care and teaching, could have been made fit to graduate to hostel placements have to remain in the institution. Adults who could have done a simple job in a sheltered setting are never taught to do so.

This failure to make the most of potential within institutions makes the gap between the beds provided at the rate of 1·3 per 1,000 and the actual estimate of 3·3 per 1,000 disastrous. The hope behind this provision of beds was that greater use of hostels and community care would offset the increase in waiting lists due to more survivals. But shortage of staff holds up the movement of patients from institutions to hostels or sheltered accommodation. And to make matters worse it also holds up the provision of such sheltered placements. There is an increasing investment in hostels, sheltered workshops and training centres – but the single biggest reason for not doing more is shortage of staff.

So officially society adopts a total-care ethic, though many individuals disagree with this. The issue has not been thrown open to public discussion, as any alternative to total care would be political dynamite. Officially, then, society believes that mentally subnormal children should be kept alive and cared for to the best of its ability. It believes that family care in the community is best, but fails to provide the support which could make this tolerable. Where it provides institutional care it provides neither the quantity nor the quality which is needed. And where it provides halfway houses the same charge holds true.

To be blunt, society is failing handicapped children and their families – and its own principles. Much is being done, but it appears that the rising tide of chronic handicap and the growing swell of consumer demand that things should be done better is proving too much for us. Our societies are affluent and we could, by a big switch of resources, give all handicapped children and adults the full support they need. On the other hand, our societies are not that affluent and there are a thousand other causes clamouring for our concern: we may prefer to turn to those and leave the defectives. And we may find cures. Which of these three possibilities we shall choose I have no idea, nor has anyone else. But is it not time that we talked about them openly?

# 8. BRAINS

The human brain is no longer considered as a kind of meta-physical blancmange called a 'mind'. Science has made it an electrochemical information-processing machine, running on spike potentials, neuro-humours, cell metabolites and the rest. By doing so it has made a profound difference to the way we picture ourselves. But it has also given us a technology of the brain and with it radically new powers to alter ourselves.

Few brain researchers doubt that these powers will increase dramatically in future. The explosion of 'inner space' is simply moving too fast on too many fronts for any other likely alternative: we are heading for a time when we can more effectively, more precisely, retune brains, alter personality, modify moods and control behaviour. As David Krech, Professor of Psychology at Berkeley, has put it,* 'I don't believe I am being melodramatic in suggesting that what our research may discover may carry with it even more serious implications than the awful, in both senses of the word, achievements of atomic physics. Let us not find ourselves in the position of being caught foolishly surprised, naïvely perplexed and touchingly full of publicly displayed guilt at what we have wrought.'

In this chapter I have selected four key areas, covering most of the ways scientists suggest we shall learn to control brain and behaviour. I trust that they will reveal which problems we shall really have to face and which are illusory. The first is a range of rather specific techniques that might allow big changes in what one can only call overall mental performance: 'intelligence', memory and so on. Second, there are surgical methods for making

* Speech to the December 1966 annual meeting of the American Association for the Advancement of Science, Berkeley, California.

highly specific changes in moods or even basic personality by implanting electrodes in the brain. Next there are the drugs – 'mood pills' – which might achieve similar ends. Finally, away from biomedicine proper, there are the social controls – rewards and punishments to 'control' behaviour.

But first, a very brief look at the overall structure of the brain. A baby is born with a three-layered brain. At the base there is the primitive brain stem which controls the involuntary functions of the body, like heart-beat and breathing. Higher up there is a collection of discrete structures known as the mid-brain. Here organs like the hypothalamus, the pituitary and the amygdala direct most of the body's hormone systems, and are the seat of drives, moods and emotions which are partly within our conscious control. At the top there is the thick envelope of the cortex, the seat of the senses, speech, voluntary muscle control, 'thinking' and memory. Forming this structure are about 10,000 million neurone cells and 100,000 million glial cells which provide a soft packing and nourishment for the neurones. At birth a baby has most of these neurone cells and by the time he is about two has his full stock. After that they cease to divide and gradually start dying: after adolescence we lose about 10,000 neurones a day. But neurone cells are not the whole picture. It is now known that each neurone can have anything from 4,000 to 10,000 nerve fibres connecting it to its neighbours, so that each of all those 10,000 million neurones can be as richly wired up to their surroundings as a telephone exchange. Vast numbers of these connections, though, are 'dead lines' that can carry no signals until they are activated and switched through to the neurone cells at their ends.

This fantastically rich wiring system is what gives the human brain its astonishing capabilities. But perhaps the most extraordinary thing about the brain is its ability *with use* to complicate and enrich its own basic structure. It is now thought that the more brains are used the more interconnecting 'wires' are laid down and the more often they are activated. And since this wiring and activating is basically a matter for the molecular machinery of brain cells, it is not only affected by how much the brain is worked; it is also intimately affected by the chemical environment of brain cells. Mentally we are far more open to change, with a far greater potential, than our grandparents ever realized.

## Performance enhancers

In 1917 an English scientist called Lashley injected rats with tiny doses of strychnine and found that they became better at learning to run through mazes. Since then, especially in the last few years, hundreds of scientists have taken up the idea, so that there is now massive evidence that drugs, a richly stimulating environment, or both, can alter the actual hardware of the brain on which learning and memory depend. Hence all the talk of 'memory pills' and 'intelligence' boosting. How much substance is there to these speculations, and what might their implications be?

Since learning basically comes down to a matter of memory – how rapidly and thoroughly new information is fixed in the brain, how permanently it is stored, and how effectively it is read out of the store – most of the experiments that have been done in this line concern the biochemical basis of memory.

It is known with fair certainty that information is fixed in the brain in at least two stages. The brain is continually bombarded by a mass of sense impressions and the vast majority of these incoming signals are rejected within a split second as not being worth storing. The more 'useful' signals, though, are put in a short-term memory store: we can usually remember a name or telephone number we have just been given though we cannot remember all that we saw three seconds ago unless it was exceptionally vivid. From this short-term store, information can either drop out (we forget the number) or it can be transferred to a permanent long-term store – once we have fixed a name or number it is relatively hard to forget it. Estimates of the time taken for this change from transient to stable memory vary from a few minutes to a few hours, with thirty minutes as a commonly quoted figure.

Before it is permanently fixed in the memory, information is thought to be electrical in nature; in other words, a neural signal circulating around through a loop of neurones and interconnecting fibres. During this time it can easily be erased by drugs, blows or electric shocks to the brain. But these erasers cannot usually wipe out the memory of events occurring more than a few hours before they are used, which very strongly suggests that the long-term store cannot be electrical but must involve some permanent, stable change in the constituents or structure of neurone cells or their interconnections.

One of the main ideas is that, in much the same kind of way that DNA can store genetic information in cells, the seat of memory is a chemical molecule of some kind. The most popular candidates are RNA (ribonucleic acid), proteins and sugar-proteins (mucoids). All of them (like DNA) are long, chain-like molecules made up of sequences of simpler 'building block' units that conceivably could carry memory traces in a coded form in the same sort of way that DNA carries the genetic code. But for good reasons the search to find out which chemical is the memory store is proving very difficult – and highly contentious.

One way of searching is actually to try to transfer memories between animals. The most sensational experiments here were done a few years ago by two American scientists, James McConnell and Allan Jacobson. They trained tiny pond animals called flatworms (planaria) to respond to simple stimuli: for example, to stretch out towards a light or curl up in a ball when given a mild shock. They then minced up the brains of these trained flatworms and fed them to untrained worms. They claimed that on this cerebral, cannibalistic diet the untrained animals now rapidly learned to respond to the same stimuli. Since then Jacobson and others have claimed the same kind of results with higher animals (and even between species) fed with RNA and proteins extracted from the brains of trained animals.

Many people (especially science journalists, because the implications are so dramatic) leaped on these experiments with speculations that we are on the brink of instant learning by injection. If RNA or proteins hold a man's memories why not extract it from professors when they die, replicate it and inject it *en masse* into children? Eventually, why not even synthesize it with specified memories built in? Instead of the classroom, a row of labelled bottles: 'Drink Me – and Know . . .' For better or worse, one can dismiss this as pure fantasy. Most reputable brain scientists insist that memory must be a far more diffuse, complex and fragile quality than transfer learning claims give it credit for, while the most outstanding thing about these claims is the difficulty anyone not enthusiastically devoted to them has in repeating any experiment and getting the same results.

The mainstream experiments on the chemistry of memory take a rather humbler approach. Broadly, there are two main types. With the first, animals are made to learn new tasks, killed, and regions of their brains studied to see if there have been any significant changes in chemistry; for example, increases of RNA or particular proteins. With the second, substances which are known to speed up or slow down brain-cell activity – particularly the synthesis of RNA and proteins – are injected into animals which are then tested to see if their ability to learn and remember is altered.

What these experiments have told scientists about the chemical mechanisms of learning and memory is not part of our story.*

*These experiments have certainly established that 'brainwork' does produce marked chemical changes – e.g. increases in RNA and proteins – and that artificially altering RNA and protein synthesis produces marked changes in learning and memory. The problem is to interpret them. Some scientists believe that they strongly suggest a chemical store. But which particular chemical? The trouble is that because RNA is the template on which cells make proteins, a rise (or fall) in RNA is bound to accompany a rise (or fall) in proteins, and vice versa. Besides this, until very recently only about 50 per cent of the proteins in brain cells had even been identified. At the time of writing, this is one of those fields of science where there are a thousand experiments and ideas for each good, solid conclusion.

Other scientists are convinced there is no chemical store as such, no memory equivalent of the geneticist's DNA. They insist that it is all to do with the synapses – the minute gaps between the ends of nerve fibres and neurones across which every signal entering or leaving a neurone has to pass. Since there are about 10 million million synapses in the brain they offer plenty of scope for storing the 10,000 million 'bits' of information that the average human is supposed to accumulate during a lifetime. The crucial point about synapses is that nerve signals do not cross them as an electrical current, like an ultra-weak lightning flash, but as a chemical flow. A nerve impulse coming down a nerve fibre triggers the release of chemical 'trans-

What does matter is that from this second type of experiment it is well established that there are many drugs that can enhance or depress learning and memory.

To take the enhancers first, it is known that a whole range of drugs – including strychnine, picrotoxin, nicotine and the amphetamines – when injected into animals can decrease the number of tries they need to learn how to run through a maze and can make them remember new tricks for longer. For example, as part of a classic series of experiments over the past few years at Berkeley (University of California), Lewis Petronovich and James McGaugh treated rats with a drug called Metrazol and found that they needed only five attempts to learn a maze; while untreated though genetically identical rats needed twenty. What is probably happening with all these drugs is a speed-up of memory fixation. In other words, new information is consolidated in the permanent store more rapidly.

From these and other experiments four crucial facts have emerged that might have interesting implications if these drugs are used on humans. First, nearly all the drugs are highly toxic. Second, one cannot just pump in more and more drugs and get

---

mitter substances' stored at the fibre tip. These diffuse rapidly across the gap to the neurone and trigger it into firing an identical impulse, which in a similar way is then passed on to a second nerve fibre, and so on. These transmitter substances are manufactured by the neurones. So too are anti-transmitter substances which break the transmitters down after each impulse to prevent them accumulating and clogging up the cell. Since the transmitter and anti-transmitter substances are proteins and are made by the RNA in the neurones it is only too easy to explain all the observed RNA and protein changes as being about changes at the synapses. For example, by injecting a performance-boosting drug that, say, makes the cell synthesize RNA more rapidly one might simply be allowing synapses to work at their optimum level, or work more frequently, or to pass weaker signals, or one might be activating previously 'dead', unused synapses. There is in fact striking evidence to support this synapse model. While the chemical store enthusiasts cheer whenever they find that brainwork produces more of this or that in the cells, the synapse-school are now pointing out rather convincingly that brainwork also produces size increases in synapses – and of course to get that the cell must make more RNA and protein.

better performances (quite apart from the fact that one would poison the animal): there seem to be optimum doses beyond which there is no further improvement. Third, the amount of improvement does depend very much on heredity. There are congenitally stupid rats and congenitally clever ones, and while both kinds can benefit from drugs the critical dose is different for each. Nevertheless, an optimum dose of, say, Metrazol can make the dullest rats perform as well as the very brightest rats when they have not had the drug. Fourth, there is some evidence that different drugs working on different genetic strains can enhance some kinds of performance but not others.

So far none of these drugs have been definitely shown to work with humans. They have, however, been tried. The most famous case was with a drug called magnesium pemoline, discovered in 1965 by A. J. Glasky and Lionel Simon at the Abbott Laboratories in Illinois, and put on the market under the trade name Cylert. The first human trials were reported in 1966 by the late Dr Ewen Cameron of the Veterans Administration Hospital, Albany, New York. He used Cylert in a double blind trial on 24 senile men, testing them with the Weschler memory scale, and various counting problems. For a week after they were given Cylert there was no improvement in their memory. Then the drug group began to pull significantly ahead of the control group: 19 of the 24 gained 10 or more points on the Weschler scale. The greatest changes took place in patients whose memory had been comparatively mildly impaired before the trial: where it was very poor to begin with there was little improvement. One patient was able to start driving again. A bridge player who had given up the game because of his failing memory was able to start playing again. A third patient began to know where he was when he woke up in the morning. A fourth remembered once more how to turn on a television set.

Despite these extraordinary results, other clinicians have completely failed to find any improvements in human learning or memory on Cylert, and in at least one case have even found that Cylert patients did worse than those given a placebo.

Nevertheless, in their less cynical moods few brain researchers doubt that one day – and perhaps quite soon – there will be reasonably non-toxic regimes that can enhance human memory and so the ability to learn, and hence to a large extent the general level of intellectual performance.

So can we now start looking at their wider implications? The answer is no. Chemistry is not all. The brain is not an isolated electrochemical machine. Chemicals may influence its chemistry but the environment can also influence its chemistry, and the brain can influence its environment which can influence its chemistry . . . Our brains are tied in to a feedback loop with many more things than a packet of pills. We have to look at the way the environment also alters brains.

Not long ago, while it was recognized that experience made the brain more efficient – made it, if you like, fuller – no one thought that it could actually alter the basic structures of the brain. It was clear that children brought up in the dull, restricted environment of, say, a bad orphanage functioned at a lower level than orphans placed in stimulating foster homes. It was also clear that if a sensorily deprived way of life went on for too long – as with illegitimate children chained for several years in an attic away from the prying eyes of neighbours – those children would always be mentally subnormal to some extent. But it was thought that all this meant only that brains function to their full potential if fully used, and not otherwise.

Now, again from work with animals, it looks as though use, stimulation, teaching, can actually alter the chemistry and the anatomy of the brain. For example, another psychology team at Berkeley, California,* has reared rats of the same genetic strain, age, sex and litter, in three different environments. A first group was reared in the normal social conditions of a laboratory rat cage. A second group was reared with each rat isolated, alone with dim lighting and little handling. The third group lived a hectic social life, housed together in well-lit busy laboratories, provided with many rat toys to play with, handled, trained, made to work and well rewarded for doing so.

* See for example: Edward L. Bennett, Marion C. Diamond, David Krech, Mark R. Rosenzweig, *Science*, 146 (1964), pp. 610–19.

The behaviour of the three groups differed in the way one would expect with deprived and busy children. But what was notable was the post-mortem differences found at fifteen weeks. The third group of rats had consistently heavier, thicker cerebral cortexes than the first rats; the blood vessels supplying the cortex were larger; the neurone cells of the cortex were larger; the total chemical activity of the neurone cells was greater, with more anti-transmitter substance in the cortex cells; and there was a thicker tangle of interconnections between the neurones.

Experiments like this have been repeated in many laboratories with many variations. Their message is the same: 'good brains' can be cultivated. The child of intelligent parents, who will tend to create a stimulating environment for him, is probably not only being given the chance to use his brain to its full potential; he is actually being given a better brain to use. Similarly, ghetto education projects like the American Headstart Programme are not only equalizing children's chances of using their potential; they are probably building potential.

The moral message for society is obvious, but hardly new. What is new are the practical consequences. Many scientists are now predicting that within a few years there will be drugs, or stimulated environment regimes, or both combined, capable of raising human 'intelligence' by up to 20 I.Q. points.

Professor Krech has painted a partly frivolous picture of what we might have to cope with, depending on how it is found that drug and environmental regimes work best, and on whom. If they worked on everybody, bright and dull, the whole I.Q. curve of the population might shift 20 points to the right. In a complex, leisured society, with its increasingly heavy demands and advantages for intelligence, this seems all to the good. But the strains of adjusting to the change, especially if the change was rapid, would be terrible. They would be even more terrible if – as is very likely – the effects were far greater on young, developing brains. Who would teach this new generation? How much would the already violent conflicts between generations be exacerbated? Feeling that their jobs and livelihoods were threatened by the advancing tide of 'unnaturally' or 'artificially' bright youngsters, the older generation might set up closed shops to keep the threat away. At the same time the new super-generation could well have an even lower respect for authority than now. And what effects would a new high-I.Q. population have on politics, the mass media, morals – in fact on everything?

Next, what if some of the drugs worked only on duller people, as those rat experiments suggest they might? The I.Q. spread, established over millennia, would bunch up at the right, high end. Already all societies fail to match jobs to brainpower: we have high-I.Q. under-educated dustmen and average-I.Q. over-privileged executives. Hewers of wood and drawers of water might be as hard to find as a seat on the board would be for the duller man who can now pull strings. But such a society would probably speed up the automation of dreary jobs and would be better equipped to face the six-day week-end.

Finally, what if the regimes only affected specific mental traits or groups of them? Mathematical ability, perhaps, or total recall, or the ability to visualize complex three-dimensional structures? The crucial concern then becomes, Who does what to whom? Who gets which ability raised? Children whose parents can afford the pills? Who decides? Parents, hucksters, the school board? On what basis? On the effectiveness of TV advertising – the pills will, after all, be manufactured and marketed by someone – or political expediency? These may be legitimate concerns. On the other hand, perhaps we should not be too awed. Already we have faced a situation in England where parents bought coaching for their children in the critical 11-plus examination, and where this injustice was met by providing optimum coaching for all, thus simply altering the competitive baseline. Similarly, is dosing a child with a special ability pill very different from getting it to love reading by providing it with a wealth of books to read, or mathematics by providing a good mathematics teacher? These are reversible indoctrinations, true; as the child grows it can voluntarily cultivate them or supplant them with other interests and abilities. But they are not totally reversible: culture and education set us on ability tracks that we can wander about in but not right away from. So, probably, with ability pills. What really matters here is whether such pills would be additive (could one enhance three abilities by taking three pills?) or whether they increase one special ability to the detriment of others. In the second case, there would be much more cause for concern. As an example, consider the child of normal intelligence who by chemical manipulation is turned into a mathematical genius *at the expense of* much of his general ability. This may make him an economically better tool for society but a more restricted, less happy human being.

While this kind of speculation may be fun, with drugs at least it ignores one vital factor: toxicity. Virtually all those rat drugs are highly toxic in even moderate doses. Even if non-toxic ones were found, medical caution about long-term side-effects – as well as cost and sheer human conservatism – would certainly delay their spread and give us more time to argue and adjust.

But this in itself could bring further problems. Because of side-effects and risks, the drugs will be used first only by the medical profession on patients who might really benefit from them – like those senile men of Dr Cameron, or the tens of thousands of institutionalized mentally subnormal. The benefits could be enormous: lives could be transformed, institutions emptied, scarce skills could be released for other work. But there would be a vast demand for jobs, houses, for leisure interests, from those who come out of institutions for the subnormal and wards for the old. In the long term we might face the prospect of younger dullards, made normal by a pill which affects their brains and not their genes, being made free to reproduce; and much more likely than not producing dull children. These children of course could be 'cured' by the pills, but the numbers needing treatment would inevitably increase.

If the drugs were mildly toxic, so that they were on prescription, or if their cost was high (which seems likely considering their promise), they could force another wedge between the 'haves' and 'have nots'. They would be snatched up eagerly by over-anxious intelligent middle-class parents – the kind of people who today believe that by enough know-how and pushing their children can be turned into princes. Such parents make it their business to know where and how to push, and they usually succeed. Theirs are the nurseries that will first look like the enriched environment laboratories. When enhancement pills become available they could well exaggerate the already yawning ability gap in our culture, while possibly turning out intellectual giants who are emotionally stunted or have personalities like vipers.

Now what about the other side of the coin – drugs that can block learning and memory, the chemical disenhancers? These are certainly here already. Many chemicals – nearly all substances that inhibit the synthesis of R NA and proteins in the brain – have been found that impair learning in animals by blocking the brain's ability to fix new memories in the permanent store. In most of them, both short-term and permanent memory are relatively unaffected: it is the step from one to the other that is blocked, so that the animal cannot learn new things.

Do we have anything to fear from the use of these drugs on man? Probably not, because as before in effective doses these drugs are very toxic. As someone has said, if you are going to block memory by inhibiting 90 per cent of protein synthesis in the brain it is not all that surprising that the unfortunate experimental animal cannot learn to run a maze – one usually finds it dead a few hours later. However, we ought to be ready for the non-toxic versions which some brain scientists foresee.

There might be real dangers in their use. A criminal, for example, might be made to forget that he had been forced to confess: a murder witness might be made to forget what he had seen . . . we can all embroider on the possibilities of this kind of chemical de-bugging. Laws could be passed making it mental assault to use them, but they would still be used, just as the so-called 'truth drugs' are misused today. What we do not know is whether a memory eraser would leave a gap in the memory which the victim could *recognize* as a gap but could not fill. If there were no gap he could never report that he had been illegally dosed, but even if there were one and he did go to the police he could be singularly unhelpful to them.

Alternatively, memory erasers could be used beneficially to wipe out experiences we would rather forget. For example, a child who had witnessed a dreadful accident or had been sexually assaulted could have the horror erased from its mind. Most psychiatrists would probably reply that the memory hole would be much more psychologically disturbing for the child than the knowledge itself and, with help, learning to 'live it through'. Also, like physical pain, mental horror is a warning device to help one avoid future situations where it might recur: the child who had no dark thoughts and fears would be a naked, innocent victim. Besides, can one really imagine a parent injecting a memory preventer into his or her child immediately after an accident, rather than comforting, supporting and explaining to it?

## Electrodes in the brain

Performance enhancers spring from our new maps of how the brain works at a molecular level. In the last few years there has been a second kind of mapping of the brain: a mapping of the specific and often minute sites in the mid-brain, the brain-stem and even the spinal cord which control the basic aspects of living, moving and behaving.

The principal technique – called electrical stimulation of the brain (E.S.B.) – is to sink pairs of hair-like electrodes into chosen spots in the brain. Electric currents passed down these can stimulate tiny areas of the brain into firing. Alternatively, stronger currents can be used to burn them out and stop them ever firing again. Much the same effect can be obtained by injecting stimulating or killing chemicals down tiny tubes called chemotrodes, or destroying cells by radioactive implants or local freezing.

E.S.B. is now a highly sophisticated technology. There are microelectrodes so fine that they can stimulate single neurones. Electrodes made of materials that are reasonably inert biologically can be left in the brain for years without apparently causing harmful reactions. Electronic miniaturization has led to implanted stimulators-cum-radio-receivers; rather than being strapped down to a chair and a mass of wires, animals (and men) can range freely and be stimulated by remote control. As a result, E.S.B. is moving very fast and is already a vast, complex subject. So the best that I can hope to do here is give a mere flavour of its range and variety.

To start at the most basic level, E.S.B. can control the previously automatic regulation of a body's economy. Heart-rate, breathing-rate, temperature regulation, local blood supply, hormone discharge, all these are under the control of specific sites in the brain – many of them in the hypothalamus – and all can be altered by E.S.B. Muscle groups can also be manipulated by stimulating the control centres in the brain to which groups of muscles are connected via the spinal cord. Animals can be made to move like electric toys; and can be made to do so indefinitely and whether they want to or not. If they resist, their 'will' can be overcome by a stronger current.

With other specific brain sites, E.S.B. can produce whole sequences of complex behaviour. For example, Professor José Delgado, of Yale University, has shown that a five-second stimulation of a particular spot in the red nucleus of a monkey brain will make the monkey stop whatever it is doing, make a face, turn its head to the right, stand on two feet, circle to the right, walk on its hind legs round its cage, climb the cage wall and return to the floor. As the stimulation ceases it will grunt deeply, stand on all fours and glare at the nearest monkey neighbour, peacefully approach it and then resume normal activity. It will repeat the entire sequence, always in exactly the same order, as often as the button is pressed. In one case this was done twenty thousand times: once a minute, day and night, for two weeks.

E.S.B. can also produce complex social interactions and alter emotional states and basic drives. Cats can be induced to fight their friends and welcome their enemies. Careful experiments have shown that the stimulated rage or friendliness is genuinely 'felt' by the animal – it is not just robot-like fulfilling of the motions. By switching stimulation back and forth between two tiny adjacent spots in the amygdala, animals can be thrown from paroxysms of hate into excesses of affection. By stimulating other areas, cats can be made to purr contentedly when they are hurt, to cower in terror at a mouse, or to become terrified of nothing at all when they are alone.

In principle, there is no reason why E.S.B. could not do the same kind of thing to men and women. What, then, are the moral and social implications of this extraordinary power to control behaviour?

Many people have leaped to the obvious conclusion that it could be terribly misused on a mass scale. In a really despotic police state They could run us all like electronic puppets, sitting at the radio consoles pushing the buttons as They please. But to control an unwilling man with electrodes you have to catch him, strap him to a chair, do a long mapping operation on his brain to locate precisely the control sites, and follow this with a long, difficult brain operation to implant the electrodes. No two brains have their critical sites in exactly the same places, so there are no short cuts to mass implantation. Even if there were, it would be very easy for the 'controlled' population to escape the button-pushers by putting themselves behind radio screens or simply pulling out their terminals.

E.S.B. will remain a research and therapeutic tool. At present most people who have had electrodes implanted in their brains have had them for the purpose of searching for and then destroying a defective brain site. In epilepsy or in Parkinson's disease, electrodes can be sunk into the suspect area of the brain and used either to record abnormal electrical activity or to produce the unwanted symptom – the convulsion of epilepsy, the tremor of Parkinson's disease. Having located it, the correct site can then be burnt out by passing a cell-killing current through the electrode pair. This is a tremendous therapeutic advance. By allowing far more precise mapping and destruction, E.S.B. is taking us away from the classical dilemma of brain surgery: 'Do I cut out (or have cut out) this bad bit of brain, at the risk of losing some good bits with it so that the patient will end up depersonalized or idiotic?'

But E.S.B. is also being used in a different, and ethically more difficult, therapeutic sense, not as a mapping tool for permanent destruction of misbehaving brain cells, but as a continuing form of treatment. By implanting electrodes and leaving them there, brain sites can be stimulated to alter aberrant behaviour or emotion: to quieten the violent psychotic, for example, or to relieve the psychic pain of the acutely anxious. At present such patients – and there are only a few hundred of them in the world – are almost all inmates of acute wards in mental hospitals. The patient is probably observed by hospital staff, and stimulated by the electrode button when he becomes unmanageable. In a few cases, though, he may manage his own button; pressing it when he feels his desperate need. The results can be dramatic. One patient, in hospital for spells of uncontrollable anger, who now has a button to control electrodes in his caudate nucleus was recently reported as saying, 'When I get mad or something, I push it and immediately feel better. It feels real good, so I just keep it up . . . that's a real good button. If I could buy one, I'd take it home with me.' So far only a very few patients do have their own buttons at home, though from reports many of them seem to be living apparently normal lives.

The next stage in E.S.B. control will probably be automatic buttons that do their own switching on and off. The technique here is to use sensing electrodes alongside the stimulators to monitor continuously electrical activity in the control site. As soon as that activity reaches the threshold which produces un-wanted behaviour or emotion, the sensors could trigger the stimulators into action. The owner of this totally automatic control system would never know when he was being stimulated. His abnormality, whether epilepsy or uncontrollable anger, would simply be smoothed out of his system and he would never again be aware of it.

Where such a procedure cured and released a mental-hospital patient, so that he felt well and could live normally despite the stigma of the electrode socket on top of his skull, it would be difficult to produce strong arguments against it. We cheer when a man can be saved from death by an implanted heart pacemaker. We accept his right to say that he is prepared to live with the extraordinariness and inconvenience rather than die. So how could we dispute a man's right to have his brain wired up if he wants to save himself from mental torment?

However, it has been suggested that if E.S.B. works for mental-hospital patients it might also work – and would be used – on criminals and the 'anti-social' generally. Although it is unlikely that a specific site could be found to suppress anti-social behaviour, it might well be possible to find sites to control compulsive stealing or the focused and apparently irrational aggression of the sex criminal.

Is this likely to happen, and if so should it be allowed to happen? My own answer to both these questions is no, because E.S.B. used in this way would only be a specific-symptom cure. And if it did get to the root cause of disturbance, then it would be a dangerous cure.

To illustrate this, imagine two men who might be typical candidates for E.S.B. in the future. Both have bouts of uncontrollable anger. Between these bouts they are as normal as anyone can be who constantly fears his own explosive emotions. The first man's destructive anger is focused on his wife: he beats her, he burns her with cigarettes, he is uncontrollably cruel and violent. His wife does not want to divorce him or bring a legal charge. She wants him cured. Instead of opting for long and uncertain psychiatric help, he goes for E.S.B., with its certainty that once the excessive spike potentials in his amygdala are found and electrodes are implanted, there will be no more anger. It sounds very good. But E.S.B. used in this way is not touching the basic defect in the marriage. The man is being electroded into living within a situation which some bit of him finds intolerable. And so his protest against the situation will almost certainly emerge in another form. Where an over-discharging amygdala has such a narrow focus as a wife, E.S.B. is unlikely to be used to replace – and certainly should not be allowed to replace – the insights and reality adjustments of psychiatric treatment.

Our second man is different in that his uncontrollable outbursts of aggressive anger have no particular focus: they just happen. An E.S.B. specialist might say that he is simply plagued by an amygdala (or whatever) whose anger threshold is too low:

it plagues him by firing off inappropriately. Inhibit the circuit by E.S.B. and the threshold will be raised. The argument again sounds good, but to what level should the threshold be raised? We have no prospect whatsoever of being able to tune E.S.B. so that a man reacts appropriately to 'genuine' (or socially acceptable) causes for anger, but not to unacceptable ones. Our patient may therefore stand by, peacefully, happily, and watch a drunk assault his wife, a thief remove his possessions, somebody hurt his baby. With his anger circuit knocked out, he could hardly function as a normal human being.

This second man is similar to the kinds of criminal on whom E.S.B. treatment might conceivably be tried: the sexual assaulter, for instance. But the man who makes sexual assaults on little girls during attacks when he 'feels funny' or 'goes all blank' has a whole complex of conscious and unconscious motivations. E.S.B. could not touch these specifically; it could only 'cure' by knocking out the whole sex drive. This would be equivalent to castration; and though castration for sexual criminals has been tried, with the intolerable frustration of impotence added to his twisted personality, such a man is all too likely to turn to child murder. So unless there is a move to 'cure' criminals by E.S.B. that rides roughshod over all psychiatric protests – and I see no likelihood of that – it looks as though E.S.B. for humans will be reserved for acute therapy for difficult organic lesions and perhaps for the comfort of mental hospital inmates – a politer version of leucotomy.

However, the most profound implications of E.S.B. will surely come from its use not for controlling behaviour but for mapping it, for understanding it. When we have better maps of the brain and everyone knows about them, there is at least a chance that we shall find it easier to control or accept our own apparently irrational behaviour. If I know, whenever I get furiously angry, that all this commotion is caused by a flurry of spike potentials in a tiny spot of my amygdala, if I can see it in my mind's eye, then I might be able to see the absurdity of my own behaviour, and calm down. After all, most women find it easier to live with their frantic irritability when they know that it arises from premenstrual tension. And their husbands, when they know, usually find it easier to make allowances. Similarly, part of the value of psycho-analysis lies in the fact that it is easier to live with difficult bits of yourself once you understand and accept them – even if they are not 'cured'.

As Dr José Delgado has put it, E.S.B. is beginning to make us realize that 'Human behaviour, happiness, good and evil are, after all, products of cerebral physiology . . .' And if this makes one feel degradingly like an electro-chemical puppet, I can only say that I prefer this feeling to the belief that I am controlled by wild devils in my head.

### 'Mood pills'

In the last fifteen years the pharmaceutical industry has poured out a stream of drugs that can affect mood. In the next fifteen years the stream will grow to a flood. Indeed, according to Dr Stanley Yolles, Director of the U.S. National Institute of Mental Health, 'The next five to ten years . . . will see a hundredfold increase in the number and types of drugs capable of affecting the mind'. For societies already almost traumatically concerned with the problems of suburban housewives stuffing 'happy pills' down their throats by the kiloton, while their children move on to richer experiences with marijuana and L.S.D., this part of our future looks tricky. Shall we be able to cope with it?

We cannot begin to look at the social impact of this kind of development in any meaningful detail until we know, first, how attitudes will change and, second, precisely what new drugs there will be – what effects they will have and (no less important) what side-effects. All we can do now is raise rather general questions; but such is the importance of the subject that even that is worth doing. Let us start with a very broad survey of the main groups of 'mind drugs' that exist, and that might be developed in the future.

At one end of the range there are the chemical-warfare weapons: the anti-riot and anti-population incapacitating agents. Euphoriants already exist that can make people so optimistic that they will do, and apparently enjoy doing, almost anything. There are depressants that induce a gloom so total that the victim does nothing at all. Cataplexogenics totally immobilize their victims because muscles no longer obey the brain's commands. Disinhibitors block or weaken the controls which normally keep behaviour on a fairly even keel, sending their victims off into excesses of talking, laughing, weeping or running. Chronoleptogenics distort the time sense so much that even concepts such as 'before and after' and 'cause and effect' lose their meaning. And there are the confusants which make people lose track of all relationships, all logic, so that the world seems completely dotty and unmanageable.

Next there are the hallucinogenic drugs. These range from a bellyful of strong drink, through a couple of marijuana cigarettes, to an almost invisible speck of d-lysergic acid diethylamide, or L.S.D. What they do largely depends on the state of mind and environment of the person taking them, and the dose. Just as the drunk can be belligerent, morose or euphoric, the perceptual distortion of an L.S.D. trip can produce an experience filled with wonder and gratification, or terror and indescribable fear – with a vast diversity of reactions in between. Similarly, just as there is a difference between the 'lift' from a couple of drinks and the D.T.s after a couple of months of heavy drinking, with L.S.D. about 20 micrograms does much the same as two or three martinis, 100 micrograms will produce a full-scale reaction in an average-weight man, and a few hundred may lead to an extreme, long-lasting psychotic state. Side-effects vary widely too, from the mild hangover after a bottle of good wine to chromosome damage from L.S.D.; so, as everyone knows, do the addictive powers of these drugs, whether true pharmacological addiction (as with alcohol or heroin) or socially induced habitation (as with marijuana). Again, side-effects and addictive properties vary enormously with individuals.

New synthetic hallucinogens will surely be produced and people will surely go on using them. The search and the need for temporary escape into Aldous Huxley's 'artificial paradises' is ingrained too deeply into human culture for it to be otherwise. But while the future will almost certainly bring hallucinogens with novel (and unpredictable) effects on the mind, the most important changes will probably be in reducing their toxicity and in understanding why people react to them so differently. Huxley's wish for highly potent, non-toxic, non-addictive 'escape' drugs without any of the physically incapacitating or damaging effects of today's could well be realized.

Spanning across the centre of the mind-drug spectrum are all the general, blanket-action agents. Barbiturates and other drugs can put the brain to sleep. Tranquillizers can calm the minds of the agitated, alarmed, tense and nervous without confusing them or putting them to sleep – unless the dose is high. Stimulants – ranging from coffee, tea and cocoa to the synthetics like benzedrine and amphetamine – can increase wakefulness, reduce fatigue and bring on a mild euphoria by tickling the brain's general 'arousal area'. Anti-depressants can not only help to fight off the black despair of chronic mental-hospital patients but lift one out of gloom or sadness, increase energy and drive, and speed up reaction times and lower inhibitions.

At first sight these drugs seem to have realized the science-fiction dream – or nightmare – of full mood control, the chemical equivalent of E.S.B. This is not quite so. Though we call euphoria or calm, depressive gloom or agitation, 'moods', and though these drugs do certainly affect these 'moods', they are hardly rapiers in their action. To put it bluntly, if a trifle unfairly, they are the synthetic equivalents of getting happy by eating a good meal in good company or avoiding the early-morning blues by keeping up one's blood sugar-level with a snack last thing at night. In other words, they are far from specific; they work by raising or lowering the threshold of action of very general systems in the brain–body complex. Because of this they suffer all the disadvantages of variable action and produce a wide range of side-effects. Tranquillizers, for example, work far more powerfully on people who are agitated or anxious than on some one lying calmly in the sun. A suitably placed electrode would make no distinctions. Stimulants are affected by previous 'mood' also, and in high doses can have devastating side-effects, including jitteriness, seriously impaired judgement and long-lasting hangovers. Many a university examiner has come across a totally blank paper submitted by an over-pepped student.

However, most researchers guess that with development these snags will be ironed out. But if they are – if the toxicity, variable effects and side-effects of these drugs are reduced – this is only another way of saying that they will become far more specific. In that case one *can* begin to talk about true mood pills.

Though defining 'mood' is largely a question of semantics, I am going to assume here that a true mood pill would have to act only on the specific sites in the brain that control fairly narrow aspects of behaviour like aggression, anger, sexual activity and so on. Do such pills exist, and are they possible? The answer to both these questions must be a tentative 'yes'. Take aggression, since the social implications of its control are so important. It is easy enough to switch aggression on or off chemically as long as one does not mind affecting many other drives and aspects of behaviour. After all, farmers have been castrating bulls for centuries and turning them into placid creatures by – though they did not know this – lowering their levels of male androgen hormones. But it is now known that such drastic measures are not necessary. For example, two Hungarian scientists have recently shown that lactating rats can be made to attack or not attack frogs placed in their cages by manipulating their hormone balance. Their hostility is destroyed by giving them oestrogen and is re-activated with a dose of hydrocortisone. What appears to be happening is that certain 'hostility circuits' in the brain are being sensitized or de-tuned by the hormones. Once sensitized, a variety of stimuli, including frustration, pain and stress, will trigger the circuits into action. But if the circuits are highly sensitized then the mere presence of something attackable will spark off an attack. There is also evidence from man that drugs like Librium act fairly specifically and selectively on the aggression–hostility centres rather than by generally depressing brain function. In much the same way that digitalis gets into all muscles but selectively acts on the muscles of the heart, it does seem that some drugs might be able to pick out very localized drive centres in the brain – and therefore control specific moods. As Dr K. E. Moyer of Pittsburgh University told a March 1968 UNESCO conference on brain research and its implications, 'The control of man's aggressive behaviour by physiological manipulation is here now. It is here whether we like it or not, and whether we consider this step to be progress or not.'

What about the implications? Can we channel the drug flood so that it adds to human progress?

Let us start with the possibilities of deliberate misuse. Again, given our present chemical-warfare armoury, mood drugs seem to add no new dimensions of horror. But what about more paternal, 'benevolent' kinds of misuse? For example, as the ghettoes start to get restless at the beginning of a long, hot summer, could the authorities start doping the water supplies with 'peace pills'? Or could they dose everyone with 'love pills' just before they parade through the streets on the morning of election day?

Though we should not relax our guard, I believe this kind of thing is extremely unlikely. Such pills would have to be extraordinarily specific and constant in their effect, or else they would create chaos. At present we just do not know whether hostility, for instance, can be suppressed without affecting intellect or ambition, initiative or creativity – or even the competitive instincts which now keep most city workers at their desks. The price of preventing riots might be a total stoppage of work. There are dose difficulties too. A dose of the right level for adults could perhaps kill a baby, as it lost all its aggressive drives towards life and perhaps even ceased to demand food. On the other hand, if (as with tranquillizers) the pre-dose mood is critical, 'peace pills' might act selectively on the most agitated, riot-ready members of the population.

But whatever the technical difficulties (or otherwise), are they socially and politically likely? If the bosses used them it is inconceivable that no one would find out what was going on. When they had found out, it is inconceivable that there would not be a major row. Would governments really be permitted to dose water supplies even 'in the general good' when there is already such violent opposition to adding fluorides to save teeth? If the government were so despotic that it overrode public opinion, would it not also take the easy short cut and quell the riots with guns instead of dope?

At the other extreme there are the more realistic prospects of a revolution in the care of mental illness. Drugs like tranquillizers and anti-depressants are largely responsible for the recent rise in discharge rates from mental hospitals. They have made conditions in these hospitals more pleasant and open. And they have kept many people from ever going into them. Outside true psychiatric practice, they have made it possible for general practitioners to maintain anxious or neurotic patients without expert help – not curing their symptoms at the root, perhaps, but letting millions of people cope with their lives. As emergency aids, they help even the most 'normal' people get through life's more intolerable crises. How much easier and more effective it is to pull oneself out of grief to present a calm face to a child returning from school by taking two pills rather than a triple whisky. Finally, of course, they are immensely valuable research tools for leading us to even more effective drugs and treatment. Once one breaks through into inventing mind-drugs one can use them to duplicate and examine normal and abnormal mind states. L.S.D. for example – one of the most reviled of modern drugs – has been an immensely valuable tool for producing 'model psychoses': much has been learned in this way about the nature of schizophrenic thought patterns, which seem to mirror the induced distortions of L.S.D. It is also a valuable tool in psychiatric therapy.

But of course such drugs are not and will not be restricted to use in medicine and psychiatry. What the 'drug problem' is about is the escape of drugs from the umbrella of medical control. As they are used in mental hospitals for severe cases they spread to the G.P.s, who use them on their most needy patients. As the news of this gets around the normal public starts demanding them and asking doctors to prescribe them. Finally, through over-prescribing, theft or even 'do-it-yourself' underground manu-facture, the drugs escape to the world at large.

The drug-taking explosion has been discussed almost into the ground in recent years, and this is certainly not the place to review all the arguments. However, if one stands back and takes a broader, long-term look towards the future, a few crucial points that have not been so widely talked about stand out.

There is a growing acceptance that the general 'mood' drugs are a normal part of life. Now that they are easily available to take the edge off unpleasant experiences or emotional states, the baseline of our tolerance to such things as anxiety is changing. We now feel that chemical help is necessary at a lower point of mental agony than before, so we take the chemical solution more easily. With the young, especially, this goes for the positive pleasure of the newer hallucinogens too: the older generation, of course, have always had their alcohol and other escapes. Because of this the future will surely only repeat the same *kinds* of problem that we face with today's drugs. Even specific mood pills will add nothing new to the situation: our familiarity with tranquillizers and so on has already prepared us for their use.

However, we have to be aware of the possibility that the escalation in the *quantity* of drugs used may lead to a change in the quality of life without anyone quite realizing what is happening. When the first freeways were built in Los Angeles they were just good, new roads. Subsequent roads were just more of the same. But now the whole character of Los Angeles depends on its freeway system, and there are those who like it and many who do not. For better or worse, a technological solution has grown into a dominant characteristic.

No society has managed to stamp out 'dangerous' drugs by declaring them illegal and then setting the police after them. If there is a strong enough demand for drugs, prohibition does not work. And neither will it work in future; indeed, it will probably work even less effectively than now. This is partly because of the growth of permissiveness: we are becoming far less sure about the rights of states or evangelical groups to interfere with private 'vices' – especially when it involves using the police. It is also partly a matter of psychology: there is a lot of evidence that illegality itself tends to attract the kind of people who want to take 'dangerous' drugs – whether from boredom or as a social protest – and even adds to their kicks. But one of the main reasons is the practical–economic one that whenever demand is high, prohibition only makes the underground supply system more rewarding and therefore more determined and efficient. In the end the only limit to such a system is not the outlets it can find but the sources of the drug. If these are natural and five thousand miles away, so that they have to be shipped in past hawk-eyed customs officers and international drug-traffic squads, the supplies *can* be partly choked off. But if – like alcohol – the drug can be made in anyone's cellar, prohibition must fail dismally. And so it will probably be with the newer synthetics. As we get more of them there are bound to be some potent ones that any chemistry graduate can synthesize – and, if he is paid enough, will. It has happened with L.S.D., with the amphetamines, and it will happen again and again in future.

The third point is that in the long run every drug finds its own level of use by a balance between effect and side-effect. Very few people drink methylated spirits, despite its kick, because it has vile after-effects, including blindness. I predict that for similar reasons L.S.D. – at least in uncontrolled doses – will decline in popularity, as it is already said to have done in the hippy centres. Addiction, of course, makes a big difference here: any addictive drug must trap a proportion of takers like a powerful magnet. But by and large, the vast majority of people do not want to throw their lives away on drugs and will monitor themselves. Although there are millions of alcoholics in the world there are hundreds of millions who drink but hardly ever get drunk.

Lastly, the key question, will we really be worse off with a cornucopia of drugs that we can dip into easily? Will we be trapped into relying heavily on chemical help to get us through life – as we have already become trapped by cars, cities, houses, television . . . ? I do not see why, or rather why it should necessarily be so much more devastating to human dignity, or so much more irreversible if it is devastating, than these other props of civilization. As long as these props are not actually addictive, do not truly run our lives, we still retain the freedom to rule them and discard them if we wish – while as long as we do not discard them but do use them we must gain an advantage. Once we get over the emotional reaction that taking little soulless pink pills to manipulate our moods is *unnatural* – and therefore somehow discreditable – the idea that we might gain becomes easier to accept. Is it really any worse to live with the help of a bottle of pills than to live at the mercy of the moods set up by our own internal biochemistry or external surroundings? Is it really worse to lighten those moods *predictably* and *effectively* with drugs than to rely on the age-old but equally unnatural panaceas like alcohol?

If we are still appalled at the prospect of a vast spread in drug-taking, then of course the challenge is not so much to prevent that spread but to change the social conditions that underpin it. The suburban housewife or the teenager who is driven to amphetamines should not be the target of scorn but of sympathy: they are only compensating themselves for the fact that they cannot cope with the boredom or stresses or frustrations or loneliness of modern life. Until we can change all this perhaps we should bless the fact that the civilizations that are advanced enough to push people into such conditions are also advanced enough to invent the modern range of mind-affecting drugs.

## Social controls over behaviour

Any society is an organized body whose primary purpose is to control the behaviour of its members. However alarming the prospect of chemical or electrode behaviour control may seem, it must be seen against this background. We all try to control each other's behaviour all the time in every single piece of social interaction throughout every day of our lives. But we do not on the whole do it very well, very efficiently or very effectively. So in some mystical way, it does not count as control – and we can go on glorifying the old ideas of free will and self-determination.

In areas of social life where at least a semblance of scientific behaviour control has become permissible, behaviour has been manipulated with considerable success through the use of rewards and punishments – or, to put it more technically, positive and negative reinforcement. The technicality is important. Rewards and punishments carry emotional connotations, but behaviour control is not an emotional matter. A positive reinforcement is quite simply that result which makes the preceding action more likely to be repeated: negative reinforcement is the opposite

One could cite thousands of examples. In education, teaching machines give positive reinforcement simply by giving the pupil permission to turn to the next stage of the programme when he gets a correct answer. Being right carries the instant reward of going on to learn more. In many mental hospitals, where because they deal with social deviants we permit overt 'behaviour control', it has been found that the traditional policy of 'tender loving care for all' regardless of behaviour is often useless in changing behaviour, and sometimes positively damaging. Patients get loving attention by producing bizarre behaviour, so the bizarre behaviour goes on. Where socially acceptable behaviour has been linked to social approval, food tokens, privilege permission and so on, and aberrant behaviour punished by the absence of these rewards, behaviour has often rapidly become more socially acceptable. For negative reinforcement at its most efficient one only has to look at strict-discipline institutions like the army. Most enlisted men spend most of their time trying to avoid punishment. Why else do they go through all those dreary rituals except for fear of what would happen if they did not?

So, with present-day society, both positive and negative social reinforcement works. Yet only positive reinforcement tends to gain easy social acceptance. Where people engage in behaviour which is positively reinforced, they tend to explain that they do it because they 'wanted to anyway'. They need not admit that their behaviour is controlled. But where they behave in a certain way because of negative reinforcement, they will say that they do it because they have to or because they 'are forced to'. Because this is such a common experience, again we usually do not see it as overt behaviour control.

Obviously there are extremely difficult issues to be faced as regards who should practise behaviour control, by what means and over whom. These difficulties have led societies to stick their heads firmly in the sand, and to pretend that behaviour control does not exist and would be nasty if it did. But once we have accepted that it does exist, does work, and is constantly practised by us all, then the problems have to be faced.

Perhaps the least disputed kind of behaviour control concerns infants and young children. We all accept that they have to learn, they have to become socialized, they have to make enormous modifications to the aimless egocentricity of their early weeks. We all accept that this must be done – but how appallingly badly we so often do it. We leave children in slum environments, with little stimulation and little verbal interaction with parents and where attempts to explore or learn are at best ignored, at worst punished. Then we send them, by law, to schools where they will be bewildered and angry. We show them up all through their schooling by a competitive system and then – when we have finished and have almost guaranteed that they will behave antisocially – we punish them for being antisocial. This is a form of behaviour control almost equivalent to putting animals behind bars. To carry the indictment further, we do now know very much better. The work on enriched environments mentioned earlier in this chapter has already taught us something of the immense capacity of the very young to learn and, in the process, to increase their own mental capacities. Work with older children, however, is now showing how, even where early stimulation has been minimal, the educational and cultural gap between slum and privileged children can be narrowed or even closed.

The methods used fall roughly into positive and negative reinforcement methods, and there are some fairly violent differences between supporters of the two approaches. Positive reinforcement usually relies on the familiar setting of the 'responsive environment' nursery classroom. In a room filled with stimuli of every kind and an atmosphere of encouragement and congratulation, the child gradually discovers, for himself and at his own pace, that he can make telephone calls, serve the orange juice, change a doll's clothes, recognize his own coat-peg by its written label, and so on. The positive reinforcement is in the approval of his teachers and in his own increasing skills.

The negative reinforcement methods are newer, less familiar.

They include the quick-fire, forced teaching of men like Dr Carl Bereiter and Siegfried Engelmann* of the University of Illinois. Engelmann scoffs: 'Discovery-oriented learning is phony learning. One child is learning and five aren't. Our approach is simply to identify the critical skills that the children lack and teach them in such a way that *every* child learns them.' The children are continually questioned, continually forced to respond, to think, to reply, to put things into words. The negative reinforcement is built in because the child must respond to get any peace: he must say 'I am eating a biscuit' before he'll get time to do so. With autistic children – where, again, because they are social deviants more 'ruthless' behaviour control is socially acceptable – more extreme methods are sometimes used. Enclosed in a booth, facing the teacher through a window, such a child may find himself in total darkness if he refuses to respond. Only speech brings back the light and the teacher's face, while further speech may produce a sweet. There are even reports of electric shocks being used instead of the darkness.

Despite partisan claims from every side, we still know very little about the efficacy of these methods. It does seem that almost any *intensive* teaching method would benefit severely deprived children who still have normal curiosity and the desire to learn. But there are serious doubts about the more extreme methods beginning to be used with autistic children. Many psychologists believe that they do little more than train the children as apes are trained to have tea parties.

So in the behaviour control of children we tend to ignore what we know, we fail to do what we know would get results, and then we gripe about the consequences. In the behaviour control of adults we are perhaps even more hypocritical. We tend to admire amateur attempts at behaviour control: marriage guidance, for example, or our numerous 'counselling' services. But we disapprove of attempts to make such work more scientific, and we keep the services so short of money that they cannot become more scientific. It is as if we said, 'Behaviour control doesn't work because we are free individuals whose behaviour cannot be mani-

*Quoted in *Life* (International edition), 24 July 1967, p. 57.

pulated. But we'll do our best to see that nobody actually tries scientifically to manipulate it, because if it can be manipulated it shouldn't be.'

Surely it is here far more than in electrodes or drugs or any *mechanical* props that our greatest potential for modifying human brains and behaviour lies. And our greatest challenges. If we all used a hundredth of the present body of knowledge about human psychology and motivations to modify behaviour constructively and sensitively, we could do so much to realize the ancient dreams of human happiness. Chemical and electrical control of the brain may bring great changes, but they must only be aids to what we would accomplish by psychological and social change. After all, we *are* rather more than electro-chemical machines.

# 9. LIFE ON THE MACHINE

Modern medicine is gaining tremendous powers to postpone death. With a growing armoury of drugs, instruments, surgical skills and machines, doctors are learning how to swop certain doom for precious survival.

Much of the time they make a good job of it. Often they restore us to a full life. But in a growing number of cases their skills are not quite miraculous enough, and they exchange our death only for a kind of half-life that may be full of pain, misery and tragedy for all concerned. When one also remembers that preserving half-lives can be extraordinarily expensive, this combination produces some of the fiercest medico-social dilemmas of our times.

We have already looked at this broad theme with the salvaging of foetuses and infants. In this chapter – indeed for much of the rest of the book – we return to it, but with the crucial difference that we are now dealing with adults. This radically changes the theme. In our culture, adults have a commitment to the idea of survival which amounts to a determination to avert death at almost any cost. Many doctors are aware that this cost may sometimes be too high in misery for the individual and his family and in resources for society. They therefore usually do their best to see that the patient understands what his future life will be like. But how can a man, faced with death, believe that *any* kind of life will not be worth living – until he has tried it? Once he has tried it, he and his doctors are committed to the course of keeping him alive and it is far more difficult for them to decide to stop than to decide not to start. This is the central dilemma that biomedical progress, when it only partially saves adult lives, is forcing us to face on quite a new scale.

Among all the life-saving techniques of medicine the most spectacularly *partially* successful ones are based on using machines to prop up or entirely take over failed organs. Long ago medicine stumbled on the fact that an enormous number of people are disabled or die when a part of them fails and that they could be saved by man-made spares. In most cases (as Figure 9.1 shows) these spares are simple bits and pieces to replace functionally simple body parts, such as an artery or joint. In other cases they are complex devices that do not save life but alleviate major disability, like an artificial hand. But since they are among the sweetest triumphs of medicine they will not concern us here.

Nor are we concerned here with a third type of mechanical spare part: machines like the artificial respirator that can take over when the brain is so damaged that it cannot keep the lungs breathing, or all the paraphernalia of the cardiac resuscitation unit. Though these produce some of the bitterest triumphs in medicine – consider, for example, the man whose heart is re-started a minute too late so that he survives with irreversible brain damage as an idiot – these are crisis machines and 'resurrection' machines. They are used after a sudden emergency on victims who are in some senses 'dead' and who obviously cannot discuss and understand their treatment. They are also machines that *have* to be used (as long as they are available) because they do give the hope of complete recovery. It is only when recovery is far from complete that they start raising really nasty dilemmas. These dilemmas, which centre on the definitions of death and the pointlessness of trying to sustain a hopeless 'life' by heroic means, we shall come to in the next chapter, on transplants.

This chapter is about two kinds of machine, the artificial kidney and the artificial heart. They are today's and tomorrow's equivalents of machines like the iron lung that with everyone's prior knowledge produce 'propped people' – humans who would have died but who now, for better or worse, can live entirely dependent on their machine. As the only alternative to transplants and as a dramatic example of the miraculous powers of medical technology they have been the focus of an enormous amount of ballyhoo. It is time we looked at their true implications.

## The kidney machine

Kidneys are blood cleaners. They keep the blood's water and 'salt' content correct and remove waste products of protein metabolism. What they filter out they discharge down the ureter to the bladder as urine. To do this they are packed with about two million tube-like filters called nephrons. Blood flows through these and, much like a tea bag in hot water, wastes diffuse out through the thin walls of the tubes when they are more concentrated in the blood than in the urine outside. If they do not do so adequately, because the kidneys are damaged by a congenital fault, an accident, or infection, their owner will die the usually slow, agonizing death of terminal uraemia.

A kidney machine can keep such a doomed person alive even if he has no kidneys at all. If his blood is circulated through an artificial kidney, where the natural tubes are replaced by sheets of synthetic membrane surrounded by suitable fluid, it can be cleaned enough to maintain life for a few days, when the process is repeated. This dramatic plug-in technique, called regular dialysis treatment, is now saving thousands who rely on their machines totally for life. It is also a vital aid for kidney transplanting in that it allows patients to wait for spare kidneys to come up, makes them fit for the operation, and saves them if their new kidney fails.

However, despite all the press pictures of smiling groups of patients, their lives saved by the miracle machine, artificial kidneys are not the immortality machines that many people seem to think. Nor is the question of why there are not enough machines to go round the only moral and social challenge they raise. In fact, at present by far the most daunting problems of kidney machines arise from the fact that they do not 'work' all that well.

One of their major practical drawbacks is that for easy access to his blood the patient has to have plastic tubes permanently joined up to a vein or an artery in an arm or leg. When he is not on the machine these are joined across outside his body by another tube called a shunt. In that way his blood can flow from

FIGURE 9.1 Spare-part man

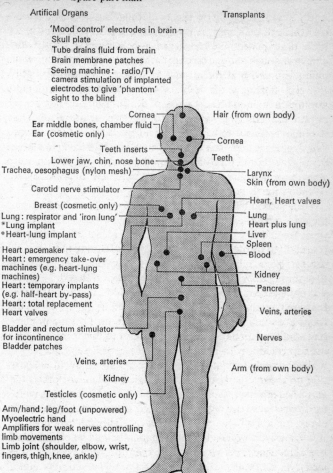

Artifical Organs

Transplants

'Mood control' electrodes in brain
Skull plate
Tube drains fluid from brain
Brain membrane patches
Seeing machine: radio/TV
camera stimulation of implanted
electrodes to give 'phantom'
sight to the blind

Cornea

Hair (from own body)

Ear middle bones, chamber fluid
Ear (cosmetic only)

Cornea

Teeth inserts

Teeth

Lower jaw, chin, nose bone

Trachea, oesophagus (nylon mesh)

Larynx
Skin (from own body)

Carotid nerve stimulator

Breast (cosmetic only)

Heart, Heart valves

Lung: respirator and 'iron lung'
*Lung implant
*Heart-lung implant

Lung
Heart plus lung
Liver
Spleen

Heart pacemaker
Heart: emergency take-over
machines (e.g. heart-lung
machines)
Heart: temporary implants
(e.g. half-heart by-pass)
Heart: total replacement
Heart valves

Blood

Kidney

Pancreas

Bladder and rectum stimulator
for incontinence
Bladder patches

Veins, arteries

Veins, arteries

Nerves

Kidney

Arm (from own body)

Testicles (cosmetic only)

Arm/hand; leg/foot (unpowered)
Myoelectric hand
Amplifiers for weak nerves controlling
limb movements
Limb joint (shoulder, elbow, wrist,
fingers, thigh, knee, ankle)

\* = not yet achieved in humans but expected soon
All other items have been achieved and are expected to have a
significant clinical impact. Trivial artificial parts such as false
teeth are not included; neither are several transplant organs
which have been achieved (whole eye) or are often talked about
(limbs from other bodies, gonads) because of severe technical
and or ethical difficulties.

artery to vein, reducing the risk of clotting. But because the tubes pass through the skin there is always a risk of infection and of the tubes being pulled out. The site must be kept scrupulously clean and dry, which means careful baths and no swimming, while injury is a constant threat, and violent exercise of the limb is frowned on. Despite the shunt, clots do form. If these break loose into the circulation there is a risk of death. If they do not, sooner or later they will clog up the vein or artery close to the tubes and the tubes will have to be moved to a new site. Until recently this had to be done every few months; now it is more likely to happen every year (in the very best units). Quite a few patients have to give up the machine – and so die – because new tube sites cannot be found for them.

Their second big practical drawback is that patients cannot be plugged in continuously. They normally go on a machine twice a week for between ten and fourteen hours at a stretch, usually overnight to disrupt life as little as possible. During this time they can eat, sleep, talk, read, play cards and when they come off in the morning they usually feel fine. But once off they are coasting downhill, accumulating poisons in their blood, and so by the time they are next due they usually feel far from healthy. To help this they have to stick rigidly to a very severe diet, on the principle that what you do not eat you do not have to filter out. Protein is usually cut down to about one ounce a day, salts have to be limited, liquids regulated extremely carefully to avoid dehydration on the one hand or 'water intoxication' on the other. Ungarnished spaghetti is the staple food in one of the most widely used regimes. Many patients find their diet a dreadfully frustrating restriction on life and often break it. This can be fatal.

Obviously, it would also help if patients could go on a machine more frequently, perhaps three or even four times a week. This idea is in fact being pushed hard, but is hardly practical or ethical as long as machines are scarce. But if or when it does become standard practice it will of course make life for the patient in some ways more trying; the strains of dialysis night are hard enough when they come 'only' twice a week. Another thing that would help is to have more efficient machines to allow shorter dialysis time, say six hours rather than ten to fourteen. Then more frequent dialysis might be less stressful; for instance, it could be done during a long evening rather than overnight away from home. Unfortunately, this might never be technically possible, since it seems that the biggest waste molecules need at least eight hours to diffuse out of the blood cells.

This brings us to the next drawback. Unless dialysis is done very skilfully with just the right dialysis times, fluids etc. for each patient, the patient may get the underdialysis syndrome. This includes chronic loss of appetite, weight loss, skin darkening, uncontrolled high blood pressure, anaemia, gouty arthritis, calcification of the bones and mental changes that can lead to suicide. With a plastic blood shunt in one's arm, suicide (and murder) is all too easy.

With these medical difficulties it is not surprising that kidney machines hardly give a guarantee of an indefinitely prolonged life. Nor is it surprising that mortality rates vary widely, depending on the skills of the doctors and nurses and the standards of their equipment. At the best centres death-rates are in fact remarkably low. As early as mid-1967 the Royal Free Hospital, London, reported that 26 patients had been on machines for 40 patient-years without a death and with all patients returned to 'normal' life. By mid-1969 many experienced units were reporting that, roughly speaking, only 10 per cent of their patients died each year. But if one looks at the pooled results from all types of unit, the results are not so good. A questionnaire to 81 European centres revealed that by May 1967, 1,163 patients had started treatment but that 32 per cent had died: 70 per cent of deaths were in the first six months. By March 1969 the number of patients who had started treatment had jumped to 4,302 but the mortality had dropped only to 30·5 per cent.* The commonest

---

*The 1967 figures are from *Dialysis and Renal Transplantation*. Proceedings of the European Dialysis and Transplant Association, 1967, David Kerr, ed. (Excerpta Medica Foundation, Amsterdam, 1968). The 1969 figures are from the proceedings of the Association's 1969 conference, published in 1970 by Excerpta Medica Foundation, ed. by W. Drukker, *et al.*

causes of death were heart failure from uncontrolled high blood pressure and failure to keep rigidly to the diet.

These astonishingly poor figures will certainly improve. They are largely a sign of rapid expansion, where almost by definition most units are struggling to learn the techniques – with their patients paying the inevitable price. Still, it is significant that these pooled results are about the same as or worse than equivalent figures for transplants in the same period. Even the best centres, with their 90 to 95 per cent chances of living a year, are doing no better than the best kidney transplant teams.

Is life 'worth' living on the machine? Most patients of course say that it is: when one is alive, however restricted that life is, what else can one say? Most dialysis specialists agree with them – obviously, or they would not be doing what they are. By and large they would agree with the words of Dr Hugh de Wardener (Charing Cross Hospital group): 'With all its limitations life on intermittent dialysis can be pleasant and fruitful. It is certainly a much more normal life than that of a paraplegic, and far less disabling than for many who suffer from rheumatoid arthritis. In spite of the restrictions ... most patients enjoy life and are at work.' *

However, some patients offered places on machines do refuse them, while others who have started intermittent dialysis abandon it. All these patients are choosing death rather than life on the machine. Many dialysis specialists insist that by doing so they should not be regarded as suicides, either legally or for insurance purposes. Looking between the lines of what they say, there is an implication that if choosing not to continue is not suicide, then life on the machine is not life in the sense that we usually understand it. With more extraordinary means of prolonging life becoming possible, this question will obviously have to be explored very carefully. When life insurance for a widow may be at stake the question is not a trivial one.

So bland encapsulated answers like de Wardener's are not

* G. E. W. Wolstenholme, *Ethics in Medical Progress*, proceedings of a symposium on transplantation and dialysis held by the C.I.B.A. Foundation, London, 1966 (J. and A. Churchill, London, 1966).

enough. If society is going to vote for the spread of kidney machines – or any other extraordinary life-supporting devices yet to appear – it must be aware of the real difficulties of these patients, and so of the extra resources it must also vote for if it is going to act humanely. Besides this, it is not often realized that the difficulties and the patient's ability to cope with them are the major factors which decide who gets a machine and who does not.

The two chief kinds of difficulty are financial and psychological. The importance of the first was brought out sharply in a study of twenty long-term dialysis patients in Newcastle-on-Tyne (England).* Fourteen of them had a drop in income because they had to give up or change their work, usually because of illness, travel time to their machine, or because they could only be fitted into day-time dialysis programmes. One man dropped £17 a week income, could no longer afford to go out and was losing touch with his friends. Income loss hit the lower socio-economic groups the hardest: twelve of the fourteen were living on savings.

Though this is not a new problem – the chronic sick often slide down into poverty – because of it some kidney-machine specialists will not accept people who might suffer in this way. Dr Stanley Shaldon, medical director of the National Kidney Centre, London, has written: 'Since the limitations imposed by cannulas [the artery–vein tubes] and anaemia preclude hard physical work, reablement is often difficult for manual labourers, who may continue their existence as a chronic burden upon their family and community. Whilst regular dialysis remains in short supply, this type of patient should be excluded . . .'†

Of course in countries without free state medicine this problem is far harder. In the U.S.A., for example, though more and more machines are now free because of government or charity money, many are still private, while insurance schemes rarely cover more than a fraction of their cost – anything from five thousand to over ten thousand pounds a year for a hospital machine. So if a patient cannot get a scarce free machine he knows he could probably buy his life – at great sacrifice – by going private. In the

* J. Goodey and J. Kelly, *Lancet*, 2 (1967), p. 147.
† *Lancet*, 1 (1968), pp. 520–23.

long run the only solution to these agonizing situations is for the state to forbid private machines, as Britain has done. But unless the state then provides as many machines as the private sector would have done, this hits hard at the rich and at the generous communities who stump up their cash for their own poor. Short of enough machines for all, there are no easy answers here.

The psychological stresses are no less difficult. Some of these arise from the patient's poor health and are not unique to kidney machines. For example, many patients lose all interest in sex; their children are often acutely anxious or frightened about a parent on the machine dying or going off to hospital at night; parents have been known to commit suicide on the machine rather than see their children watching them die slowly and without dignity. Other stresses, very interestingly, seem to depend on the emotional 'honesty' of the doctors in the kidney machine centre. If they jolly their patients along, try to create a good 'team spirit' on dialysis night, and stress how good the treatment is, patients usually have far fewer emotional and psychological reactions and suicides are rare. But if the doctors frankly stress that this is an experimental, risky procedure with no guarantee of long-term survival, patients react very differently. At the Georgetown University Hospital, Washington, where the doctors take this approach, of eleven patients in one study two withdrew from treatment (and so died), one committed suicide even more directly and two insisted on transplants (in 1965, when they were far less successful than now) because their kidney machines were 'leading them nowhere'. Two patients had schizophrenic-like episodes, one had a severe psychotic depression and all but one of the remainder had some kind of 'severe neurotic depressive reaction' to the stress of being on the machine.* This was despite

---

*E. J. Shea, et al., 'Hemodialysis for Chronic Renal Failure: Part 4, Psychological Considerations'. This is one of a set of four papers by the Georgetown group, the others being (1) technical considerations, (2) biochemical and clinical aspects, (3) medical, moral and ethical and socio-economic problems. This third paper in the set is extraordinary in that it poses no less than 85 questions for medicine and society to answer. All from Annals of Internal Medicine, vol. 62 (1965), no. 3.

detailed psychiatric evaluation of all patients before and during treatment. The obvious moral is to choose cheerful doctors. But, more seriously, how much should the success of medicine rely on emotional manipulation of the patient and 'con them to be kind' policies?

Better machines could clearly reduce these stresses by letting the doctors become honestly optimistic. But will they ever reduce the basic psychological tensions inevitable in kidney machine treatment – especially the strain of being totally dependent for life on a machine and on a flimsy pair of tubes in one's arm? Many patients are acutely anxious when their blood-shunt is removed to connect them or to take them off the machine; they watch with frightened expressions, or cover their faces so as not to see, or stare rigidly at the ceiling. Insomnia on the night before 'machine night' is common. There is frequently a strong emotional transference to the doctors and nurses, similar to that of a patient to his psychiatrist. Many patients even have a transference reaction to their machine, mixed with a feeling that it threatens their identity, that they are becoming mechanized. The Georgetown group, for example, had an adolescent patient who before he died drew people as cylinders, the shape of his kidney machine. While a kind of 'We're all in it together, like London in the blitz' atmosphere on dialysis night can help to damp these reactions down it can also exaggerate them: when one patient is having trouble everyone else realizes that they are all walking the same tightrope. Dialysis night, with all its stresses, is often dreaded; but the patient has to face it again and again and again.

Given time, many patients apparently adapt to the extraordinariness of their condition. Given much more time, our whole culture may adapt at a deep emotional-psychological level to the idea of being plugged into machines for life. In the meantime is it not essential to assume the contrary, and to insist (as some dialysis units now do) that psychiatrists have as big a role to play as doctors in the treatment of plug-in patients? The psychological management of life on the machine could well become the major problem, calling for a massive input of scarce skills sorely needed elsewhere.

Now let us turn from the 'failure' to the 'success' problems. If despite these difficulties kidney machines work well enough for most dying patients to clamour for them, can we meet the challenge presented by their success as a routine treatment? Above all, could there be enough machines for all? What would providing enough involve? And while there are not enough, how about that nasty business of patient selection? Ironically, as we look at the answers to these questions we shall come back to facing the 'failure' problems again in new disguises.

As most people know, kidney machines are expensive in money and manpower. Exact costs are impossible to give because standards and assumptions vary widely, but a reasonable estimate for building a well-equipped unit in a city hospital in Britain is as follows: *

*Capital cost* for 10-bed unit treating 30 patients
(each in for two nights, unit works 6-night week)

| | |
|---|---|
| Buildings | £32,000 |
| Machines, monitors, bed weigher, dialysis fluid supply system (£4,000 to £15,000 depending on standards), bed, pumps, piping, etc. | £24,000 to £35,000 |
| | £56,000 to £67,000 |
| Per patient: | £2,000 (about) |

*Running costs per year*

| | |
|---|---|
| Staff salaries | £18,000 |
| Disposable kidney machine coils, drugs, depreciation | £24,000 |
| | £42,000 |
| Per patient: | £1,400 |

*Manpower*

| | |
|---|---|
| 8 consultant and 11 registrar half-day sessions; 2 sisters, 9 nurses, 4 technicians and assistants full-time | 17 full-time staff |
| Per patient: | ½ person full-time |

*These (and subsequent) cost figures are condensed from estimates by Dr David Kerr in 'Symposium No. 9: The Cost of Life', *Proceedings of the Royal Society of Medicine*, 60 (1967), pp. 1195–1246. Other estimates are £2,000 capital plus £2,100 running (Dr Shaldon) and £500 capital by

How many need this kind of money to keep them alive? In Britain about 7,000 people die each year from kidney failure. About half of them are very low-priority candidates for kidney machines (or transplants) because they have other serious disorders, especially heart disease. Children under fifteen are almost invariably rejected (though they might get a transplant) because dialysis stunts their growth and prevents them reaching puberty. This leaves the following groups of potential prime candidates: age 15–44, about 1,000; age 45–54, about 1,300; age 55–64, about 1,800; a total of just over 4,000 a year. However, nearly all units reject the over-55 (and some the over-50) as long as machines for the others are scarce. So the number of prime *medical* candidates is about 2,300 a year.

Now we can estimate the cost of providing machines for all. Imagine that Britain suddenly launched an instant crash programme to get all those 2,300 new patients each year on to a machine. By the end of the first year the country would have had to have found £4·6 million in capital costs and would have a running cost of £3·2 million a year. What happens after that depends on how many patients die on the machine: clearly, if all died within months the next year's batch of 2,300 could take their empty places. Assuming the best 90 per cent one-year survival-rates (in other words, every patient has a roughly fifty-fifty chance of surviving five years) then each year there would be nearly the same £4·6 million capital cost to find, while the running costs would *increase* by nearly £3 million a year. In the fifth year, for instance, running costs would be £13 million for 9,200 patients.

converting existing wards – if they can be spared – plus 'rather less than' £1,500 running (Dr de Wardener). As for manpower, a 1966 survey of 81 European units showed on average 6 nurses and 3 doctors full-time for each 10 patients, a staff/patient ratio of nearly one. Costs could fall (see later) but could just as easily rise. For example, Kerr's building costs are reckoned at £10 per square foot for 3,200 square feet. In late 1967 the Ministry of Health, fearful of the risk of hepatitis, recommended 10-bed units to have 5,400 square feet. This pushes up the capital cost per patient by a third to £2,700. Similarly, if equipment costs fall doctors are also likely to want new ranges of equipment to achieve a higher standard of treatment.

These costs would go on accumulating but at a steadily decreasing rate, until eventually, when there were so many patients on machines that the number who died each year equalled the 2,300 new candidates, capital costs would drop to zero and running costs reach a steady level. When and at what cost this will be no one can say: kidney disease patterns will change, success-rates could improve, costs will rise or fall, selection criteria might alter. Working from today's figures Dr Kerr estimates that after twenty years about £30 million a year will be spent in direct running costs and over 10,000 staff will be treating 23,000 patients. In contrast, if all the new patients got immediate transplants at £1,000 a time the yearly bill would be more like £2·5 million.

Are these costs too high? Must there always be a 'machine gap'? Not necessarily: similar medical-care costs for large numbers have already been willingly accepted by society. For example, Dr de Wardener has calculated that in Britain in 1964 there were 30,000 tuberculosis patients in hospital or sanatorium beds costing (at 1966 figures) about £52 million, or just over £1,500 a head. However, since then medicine has had to become more cost-conscious as its power to do things has outstripped its resources for doing them and so (with the medical hazards of rushing too fast also well in mind) no state has committed itself to more than a slow, steady expansion of kidney machines. In Britain in 1967 about 200 people, and in 1969 just over 600 people (out of 2,000 prime candidates) were on a kidney machine. This was the *result* of a major policy decision to expand the programme. At the same time there were about 100 kidney transplants in the year (to be increased over the next few years to 600 a year).

This brings us to the fierce dilemmas surrounding the selection – and rejection – of patients. If for years to come only a proportion of candidates can get a machine (or a transplant), who should live, who be allowed to die? And who should choose?

In the last few years the mass media have grossly over-emphasized and distorted these problems. Hypnotized by the life-and-death drama of doctors making god-like decisions on human life; tuned to the possibility that politicians might be chosen rather than postmen, dukes rather than dustmen, working hard on the 'Isn't it a scandal there aren't more machines?' theme, they have totally missed the main point that by and large it is *the patients who select themselves* – and not by their social standing or influence either.

First of all, it is often forgotten that there is only a need for selection when a new machine is purchased or a machine falls free because a patient dies.* If ideal candidates have died in the meantime that is too bad: though it makes selection into something of a game of chance, the alternative of giving them the places of 'less worthy' patients already undergoing treatment is unthinkable. Second, any one doctor is not making these choices every day: with a thirty-patient unit and a 10 per cent annual mortality he has to select new candidates only about three times a year. So how does he do it? It might seem fairest to choose on a queue system, but unfortunately anyone at the head of the queue will have had kidney disease for longer than the newcomers and will probably be in a worse medical condition. Just because machines *are* scarce and *must* be used on patients with the greatest chance (a) of surviving a long time and (b) of surviving with least physical or emotional distress, selection must first of all be on medical and psychological grounds. Most doctors start narrowing the choice by only considering candidates between puberty and 55 who are not suffering from any additional disease that cannot be easily controlled and who are 'in their right mind and likely to be co-operative' (de Wardener). We will come to these psychological factors in a moment, because they are crucial. But the point here is that on medical grounds alone (with or without psychological ones) doctors can nearly always pick out a natural winner from those who are still alive without any need to start considering difficult social criteria. If they cannot it is rarely difficult to tighten

* If a patient frees a place by having a *transplant* a machine must be kept free for him – or at least the unit must keep some spare machines unused – in case of rejection. In the past, transplant patients have often died because there were no spare machines for them to fall back on. Doctors are now insisting that they must get top priority, even if this does mean denying a place to a new customer, on the grounds that once any treatment is begun it must be followed through.

the medical criteria and then have them – and them only – assessed by colleagues who know nothing about the rest of the patient's background.

Real difficulty only starts in the very rare instances where there are two or more natural winners. Then the doctor has to judge human worth. To ease this dilemma, the dialysis pioneer Dr Belding Scribner – who used to say that he would always favour a church-goer and always turn down prostitutes – started his famous layman's panel at Seattle (U.S.A.) in the early 1960s to spread the responsibility. At first this idea was loudly hailed as a sound democratic move, if a rather extraordinary one, and a few units have since followed suit. Today most doctors reject it out of hand, and for good reason. A small panel is bound to have the built-in biases of its members, with the policeman favouring pillars of the community, and so on. Who selects the selectors? A large panel, on the other hand, though it would smooth out these biases, becomes totally unworkable. But, above all, there is the insuperable problem that someone who knows all the social and *medical* circumstances of each candidate has to brief the panel and will therefore almost certainly feed them his own biases. That someone, of course, can only be the doctor the panel is designed to replace.

More recently several people have gone to the other extreme and proposed selection by lottery. Though this would eliminate all subjective judgement values of 'worth', it is nevertheless the most unfair system imaginable. To allow dice to choose between people would not only condemn some very 'worthy' people at the expense of others; it would be a kind of treason against human compassion and responsibility. The public might accept it as the practically fairest possible method – the Dice of Fate, scrupulously impartial and unbribable – but I would hope not.

In the end, does not the problem boil down to this? Any decisions on 'worth' must be arbitrary and must vary depending on who is making them. Yet someone has to make them. Given that, why is it such a terrible thing to leave them to the doctor, with the help of his colleagues? Their decisions may differ from yours and mine but are no more (or less) arbitrary, while they have the supreme advantage of knowing all the circumstances of their patients more fully than any other potential selector – if, that is, they care to take the trouble (which not all doctors do). If society objects to this situation, then is it really so difficult to arrive by public debate at less arbitrary guidelines for the doctors to use? For my part, I should like to see these guidelines built round the concept of human, not intellectual or economic loss. The death of a parent of a young family, no matter how 'inadequate', is a more damaging loss than the death of a great poet or prime minister, even though the latter may be mourned (at a lower level) by millions of times as many people. Besides, as hardly anyone seems to have pointed out, the poet or prime minister – or anyone else chosen on his potential value to society rather than to his family – is likely to lose his potential when he goes on a machine. Do you save a prime minister so that he can become a backbencher?

In fact it is not this kind of selection dilemma that really worries most kidney-machine specialists. What worries them far more is that one day there *will* be enough machines for all, so that they will have to start plugging in the people they reject today because they are psychologically or socially unsuitable. Far from relieving the problem of selection, more machines will only make it worse – unless they arrive with vast social security and psychological support services.

Here, for example, is Dr David Kerr on his fears for the future. After citing the case of a nineteen-year-old girl who was admitted to hospital nine times for a total period of eight months during the seventeen months she was on a kidney machine, he writes, 'This girl is a disastrous failure on intermittent dialysis; she was badly selected, lacking the intelligence and willpower to adhere to any strict diet . . . It is our impression that *nearly half* the patients with renal [kidney] failure attending our clinic would be equally unsuitable candidates for this therapy and that our troubles would really begin if we ever had facilities to meet all comers' (my italics). At present one of his main selection criteria is a 'trial by ordeal' on the strict diet. Candidates are started on it several months before they might get a machine and if they cannot stick to it they are probably refused. Other doctors stress that psychological stability, emotional maturity and a determination to make the treatment work are essential qualities. What is going to happen when there are machines for people who lack them?

These qualities are even more crucial when the patient has a machine in his own home. It is to home dialysis that many people are looking as one of the main ways of providing more machines, and in fact a major swing to machines in the home is already under way. One very strong reason is the severe risk of an outbreak of serum hepatitis from the large quantities of 'loose' blood in hospital kidney units. This disease is often fatal and has already killed many nurses, doctors and patients in kidney units. Large hospital units may even be abandoned because of it, and replaced by small clinics concentrating on training patients to use their own machines at home.

The savings on home machines are considerable. They cost more to build, mainly because for a patient or relative to manage them easily and safely, they have to be foolproof, designed for easy cleaning, automated and fitted with automatic alarms so that everyone in the family can sleep on dialysis night. But by cutting out hospital overheads and staff the running costs are low. One 1967 costing for a pilot scheme in Britain has been made by Dr Shaldon:

| | |
|---|---:|
| Capital equipment | £3,500 |
| Direct running costs: | |
| Throw-away membrane coils | 500 |
| Drugs | 100 |
| Depreciation | 350 |
| Staff salaries (servicing, hospital check-ups, etc.) | 200 |
| | £1,150 |

Equipment costs are dropping fairly rapidly while machines are becoming even easier to run: for example, one recent model sells for £1,200, only needs to be plugged into the mains supply and a cold water tap and needs no supervision. For the patient the main advantages are that he can go on the machine when he likes and more frequently (many are on for three nights a week), he does not have to leave his familiar surroundings or children or face the often anxiety-promoting atmosphere of hospital, and there is a lower risk of infection (particularly hepatitis).

So the *practical* advantages of home machines are enormous.

But how many patients and families can cope with the fearful stress and responsibility of this 'life in our hands' treatment? Dr Scribner, the American pioneer of home machines, long ago predicted that they will make patient selection 'natural' rather than arbitrary: 'Those patients who can learn to treat themselves will survive – those who cannot will die.'

Doctors obviously cannot accept this calmly, but the only way they can do something about it is to make patient selection doubly stringent. As Dr Shaldon, the leading British enthusiast for home dialysis, has put it, apart from medical criteria, 'the requirements are emotional stoicism, self-control and average intelligence'. Brighter patients are often very hard to train because they are reluctant to accept the unnatural aspects of the procedure and are often extremely anxious; patients with below-average intelligence are often unreliable and irresponsible enough to endanger their own lives. On the earlier, complex machines mechanical aptitude was critically important; even on the newer 'de-bugged' models it is highly desirable because it saves service calls. There has to be a stable home life, a dedicated and competent relative, a spare room, running water and a telephone to summon aid in emergencies.

If a patient can match all these requirements, the prospects are said to be far better than for hospital treatment: survival times are higher, complications necessitating a hospital visit or readmission are fewer and general health is better. But though there are a few medical reasons for this, as mentioned above, the main reason seems to boil down to patient selection. The person who goes on a home machine has *already survived* the critical first few months in hospital and he has been carefully chosen as someone who is likely to continue surviving at home.

And how many might be chosen, might match the stringent

requirements for home machines? Dr Shaldon has estimated 10 per cent of all candidates, and no one else seems to guess more than one in five. Home machines will be an answer only for a very few.

In the distant future technology may by-pass these dilemmas. If it could produce a cheap, effective, totally trustworthy, continuously operating, fully portable kidney machine that a patient could wear strapped to his body, it would make life on an artificial kidney medically, psychologically, socially and financially easier to bear. If it could be totally implanted in the body, so much the better: out of sight it would be that much more out of mind and there would be no external tubes to threaten life. It would be foolish to say that these developments will not come: biochemistry is only in its infancy. But as we shall see now with the example of the heart, the road to implanted machines is not going to be as easy as some optimists suppose. Meanwhile, the future of the present generation of kidney machines seems to look a lot more rather than less difficult.

## The artificial heart

The human heart is an extraordinary organ. The size of a man's fist, weighing 10 ounces, it pulses 40 million times a year to drive nearly a million gallons of blood through 60,000 miles of tubing. It generates roughly 20 watts to do this, and makes this energy itself. It can rapidly alter its beat between 60 and 200 strokes a minute, its output from 2 to 12 litres a minute, in response, for instance, to sleep, anger, frights, listening to music, kisses or playing tennis. It is self-repairing and needs no maintenance. It is just the right size for the body it grew up in and is beautifully adapted to it. It is designed to live in and pump an incredibly fragile yet reactive fluid without damaging it or being damaged by it. But it often fails.

The plan to replace this pump with a totally implanted machine is one of the most daunting challenges that technology has ever set itself. Except for two differences, one might compare it in its audacity, its need for brilliant effort and its possible benefits to the achievement of turning the Wrights' first primitive aeroplane into a jet airliner. But the differences are crucial. The first is that the time scale will be vastly compressed: already scores of animals have had their hearts totally replaced by machines (though not with much success), already scores of laboratories and research groups have developed their prototype hearts to a stage where they are nearly ready for use, and already sober estimates are putting the date for the mass insertion of mechanical hearts as somewhere in the mid- or late-1970s.

The second difference is that while you do not have to fly until they have made flying really convenient, with artificial hearts you fly or die. When for millions an artificial heart (or transplant) is the only alternative to death, how shall we cope with the agonizing development period before the obstacles are ironed out, when plastic hearts will swop death for a chance of a possibly very severely restricted life? As one of the U.S. aerospace firms commissioned to study the feasibility of artificial hearts has put it: 'All we can demand of a permanent artificial heart device is that it compensates for its user's cardiac dysfunction with only tolerable limits on his comfort and freedom of action. Any requirements beyond this represent luxuries and should be made with caution lest they prevent attainment of the basic necessity.' * And then, when we do get to the luxury, jet-liner stages, will it ever be possible to close the gap between supply and demand – a gap that could make the kidney machine scarcity look trivial by comparison?

Broadly speaking, it looks as though there will be six major steps in the development of artificial hearts (see Figure 9.2). These will not necessarily come one after the other; there may be parallel advances up several steps at the same time. But each will involve fairly big technical leaps and each will make mechanical

---

* This quotation and virtually all the factual information in this section are from the massive four-volume report commissioned by the U.S. National Institutes of Health from six major aerospace firms and research laboratories, called 'Six Studies Basic to Consideration of the Artificial Heart Program', co-ordinated and published in October 1966 by Hittman Associates, Inc., Baltimore, Maryland.

FIGURE 9.2  Six steps to the artificial heart

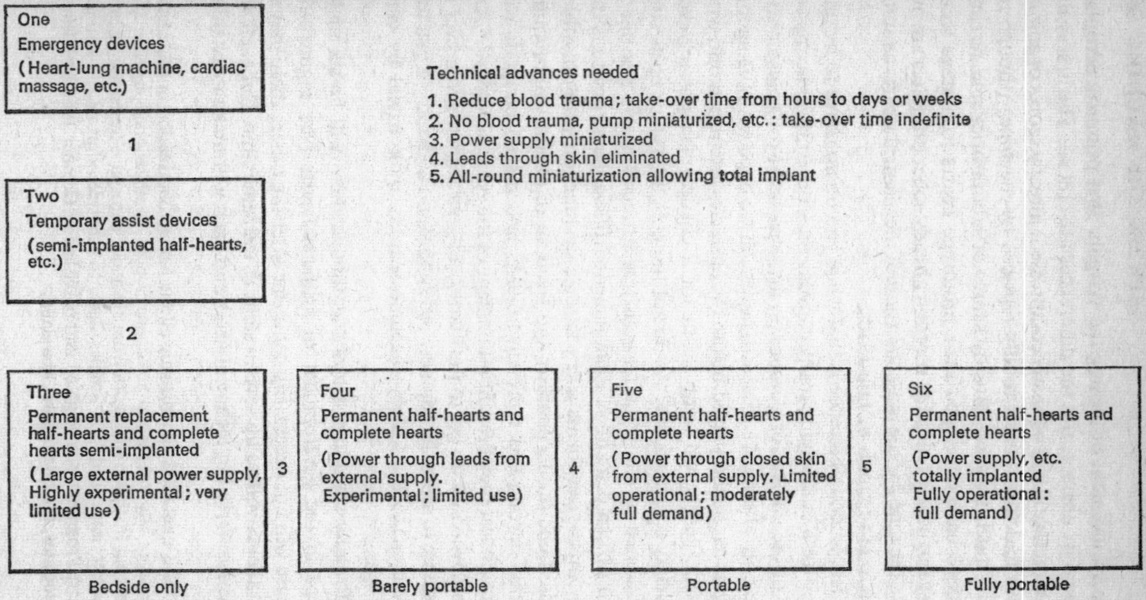

**One**
Emergency devices
(Heart-lung machine, cardiac massage, etc.)

**1**

**Two**
Temporary assist devices (semi-implanted half-hearts, etc.)

**2**

Technical advances needed

1. Reduce blood trauma; take-over time from hours to days or weeks
2. No blood trauma, pump miniaturized, etc.; take-over time indefinite
3. Power supply miniaturized
4. Leads through skin eliminated
5. All-round miniaturization allowing total implant

**Three**
Permanent replacement half-hearts and complete hearts semi-implanted
(Large external power supply. Highly experimental; very limited use)

Bedside only

**3**

**Four**
Permanent half-hearts and complete hearts
(Power through leads from external supply. Experimental; limited use)

Barely portable

**4**

**Five**
Permanent half-hearts and complete hearts
(Power through closed skin from external supply. Limited operational; moderately full demand)

Portable

**5**

**Six**
Permanent half-hearts and complete hearts
(Power supply, etc. totally implanted Fully operational; full demand)

Fully portable

hearts that much more acceptable and so that much more in demand.

We have already taken the first step. Heart-lung machines can take over from the heart entirely, allowing surgeons to operate on the inside of a virtually dry, still heart. Open or closed chest massage can also support a failed heart to give it a chance of recovering on its own. But none of these *emergency devices* can work for more than a few hours at most.

The next step is to increase those few hours to days or weeks. An enormous number of people could be given new strength and years of life if only their hearts could be rested for a time – particularly the left side that works hardest and most often fails, since it has to pump blood round the whole body rather than merely the lung circulation. What they need are *temporary assist devices* – either half-hearts that are implanted and then removed to leave the restored natural heart to carry on, or heart-lung machines outside the body that can take over the complete heart and lung function for a time.

This step has also been taken. Mechanical half-hearts have been stitched into a handful of human patients, notably by the American surgeons Dr Michael deBakey in 1963 and Dr Adrian Kantrowitz in 1966. Though some patients recovered sufficiently to have the machine removed, the record was so dismal that human trials were stopped. They were restarted in April 1969 when Dr Denton Cooley for the first time replaced a man's heart with a whole-heart device, hoping to buy time until a natural heart donor became available. This occurred after 40 hours, the machine was replaced by the donor heart, but the patient died a day later.

Animal trials, however, are going ahead apace. Scores of animals, especially calves, because their hearts are similar in size to man's, have been fitted with half-heart machines (and with whole-heart replacements). The vast majority of them have died within a very few days.

The overwhelmingly important reason for these failures is blood clotting and blood destruction. These are serious enough hazards even with artificial heart valves in humans, let alone half-

or whole-hearts, and they are the limit which prevents today's heart-lung machines from being useful for more than a few hours on any patient.

The chief culprits are poor pump and valve design and, especially, poor materials. Despite intensive development, man-made pumps and valves are too violent for blood: they create turbulence, stagnant regions, too rapid blood flow in places, and actually squash the blood – all of which adds up to clots and blood destruction. Similarly, though more subtly, with materials. Apart from the metal cages of the ball-and-cage valves usually employed, artificial hearts are almost entirely plastic. Though the synthetics like sheet Silastic (a silicone rubber), Teflon and woven-mesh Dacron that are used are among the most inert of all materials, when put in moving blood all kinds of things happen. For example, within a second they are covered by a thin layer of protein (fibrin) which, if they are at all rough, quickly builds up to a clot that might either break off or grow to block the flow. Even if they are ultra-smooth, clots are likely. Though anti-coagulants can lower the clot hazard they cannot do anything about other chemical changes which damage the blood and the materials. The ultimate solution must be to make the surfaces totally compatible with blood.

As with design, research is moving fast here. All sorts of schemes are being followed up, but broadly they all try to repel the negatively charged blood components away from the plastic surfaces by charging them too – perhaps with a battery but more probably by coating them with thin layers of substances such as the anti-coagulant heparin. The main problem is to find electrically suitable coatings that will stay on as the plastic heart flexes. Another approach is to coat the surfaces with velours in the hope that they will quickly build up a natural tissue lining.

Given the speed of development, many experts predict that by the early 1970s blood damage will be adequately conquered and that temporary heart-support devices will begin to be used widely. Initially they will probably be half-heart pumps to by-pass and rest the left or the right side only. But if it turns out that taking over one half of the heart merely adds to the load of the other, often diseased half, then total heart by-passes will supersede them. Several whole-heart pumps are well along the development pipe-line and might possibly be used as by-passes. Alternatively, they may be superseded by semi-implanted or external miniaturized heart-lung machines that take over the complete heart and lung function for several days or weeks. The prototype of these devices, developed by Dr C. W. Lillehei of Cornell University, was first used on human patients in 1968. It is based on ultra-thin Silastic membranes through which oxygen and carbon dioxide can pass. The membranes are folded up into a multi-layer sand-wich to make a blood-membrane-oxygen-membrane-blood-etc. stack. This stack copies the lung function by aerating the blood through the membranes; by moving the stack like a concertina the blood can be made to pulse through the machine and back to the body, thus also copying the heart's pumping action.*

With these temporary assistance devices it will not matter much that patients will be tied by wires or tubes to big power and control consoles by their beds: they have to stay in bed, preferably immobile, anyway. Nor will it matter much if their mechanical hearts are big, heavy, vibrate a bit or beat only at a constant rate: they can stick out of the chest and as long as they are medically satisfactory the patient will not have to put up with them for long.

---

* This device could just conceivably lead to a fully implanted heart-lung machine for permanent heart and lung replacement. Lillehei has implanted devices of this kind in dogs in the form of an $8 \times 4 \times 4$-inch box and achieved survival times of 24 hours. A human would need three such boxes, so there must be a good deal of miniaturization – quite apart from solving such problems as blood damage, miniaturization of power supplies, etc. – before it is feasible. Others are trying to develop implanted artificial lungs only, for example, by packing a tube with fine tubes of silicone rubber, passing oxygen down them and blood down the space between them, and connecting the whole system to the pulmonary artery and left auricle of the heart. This system has been implanted in animals but only to add a third lung. It is safe to predict that implant lungs or heart-lungs are many years off.

With the heart, lungs and kidneys, the fourth major replacement organ is the liver. Since the liver manufactures some 5,000 chemicals the prospect of copying it artificially is almost infinitely remote. For the time being kidney machines and heart machines are all we have to worry about.

He will either die with the machine in his chest or he will recover enough for it to be gradually shut down and then removed. But for the next steps these limitations are crucial.

The third step is the *permanent replacement* heart: you lose your own heart for ever and get a plastic one instead. This step could in fact come very soon after the previous one, for if temporary assistance devices are going to cause so little blood damage that they actually allow patients to get better with them over a few weeks – and that, after all, is their point – it is possibly only a short step to their causing so very little such damage, if any, that a patient could live with a device indefinitely.

If this step does come quickly it could produce some very nasty problems, for it would probably come before the power supplies and controls for the heart could be made portable. So there is the prospect that patients will have permanent hearts implanted but with immobilizing bedside power-control consoles. Though they will be far better off than the few pathetic people who live their lives in iron lungs, or than quadreplegics, or than the millions who are totally bedridden by a dozen crippling ailments, they will in a real sense become victims of their machines and will know this before they consent to the operation. In some tragic cases there will be no chance to consent: some patients fitted with a *temporary* heart will not be helped enough by it for the mechanical heart to be removed without killing them but will be helped enough to go on living. These people will become unintentionally trapped by their treatment.

Though everyone working on man-made hearts seems to agree that this stage must be 'purely experimental' and limited to as few patients as possible, should it be allowed to happen at all? Should society insist, now, that permanent heart replacement is not on until the power supplies are portable and there is at least some hope of rehabilitation? In a fundamental sense, society has no right to make any such strictures at all. With this or *any* other kind of heroic medical procedure – including, let it be said, heart transplants – no one but the patient and his doctors have any basic right whatsoever to question whether what is done is 'right' or not. The only thing one can do is to judge each individual case on its own, for all that matters is whether each patient gave his free and informed consent. If he knows as much as it is possible for him to understand of what his prospects are and what the alternatives are, then consent is his affair – and no one else's. This is not to say that all heroic operations are based on full consent, but that no one should dare damn them unless they know that in this case or that case there was not full consent. Having said that, though, there is still the awkward question of whether it is 'right' to offer patients such agonizing choices, especially in a culture where death is an unthinkable alternative and most people would opt for 'life' of any kind – perhaps bitterly to regret their decision later. And there is still the question of whether scarce, expensive skills should be used in this way when they could be better used elsewhere.

We shall come back to these questions later. Meanwhile, what hope is there of climbing to the next step, where one can carry an artificial heart plus its power supplies around?

At present literally scores of different systems for driving artificial hearts are being developed. Each of them has to solve three main problems. They have to provide a long-lasting *energy source*; a *motor* of some kind to convert this energy into movement; and a *control* system that couples this movement to the heart pump itself, mainly to damp out any fluctuations, but ideally to allow changes in pulse-rate, etc. And none of the three devices must be too heavy, large or vibrant, extravagant in waste heat, or expensive.

Of the three problems, control seems to be the least difficult, because lowish standards may be acceptable. Thousands of people are now living with their natural hearts driven by pacemakers at a fixed beat rate, and although they cannot run up mountains and do not get a racing pulse when they kiss they can adapt to a 'tolerably' wide range of exercise demands and emotional states. What happens is that the diameters of the blood vessels round the body alter in response to these demands and states. The heart senses this by small pressure-changes in the blood returning on the vein side and alters the volume of blood it pumps – but not the beat rate – accordingly. Building this limited adaptability into a mechanical heart is not thought to be difficult, but whether it will give more or less tolerable limitations on life than pacemakers now do, no one knows. To go further and tie mechanical hearts into all the normal body signals that make it such a remarkably adaptable machine – from direct nerve control to hormone signals in the bloodstream – could be exceptionally difficult.

Given this limit, 'good' hearts will depend on the development of motors and power supplies. At present, out of all the possible systems that are being explored it looks as though only a handful are likely to be feasible.

The two most likely to be ready first both work from electricity. In one it is a rotating electric motor that drives the heart, in the other a piezo-electric motor (based on the property of some crystals to vibrate when a current is passed through them). Both at the moment have serious weight drawbacks. Rotating motors, for example, can be made efficient enough only by running them at high speeds – but that means incorporating a reduction gearbox. The 'crystal motors' give out very little power, necessitating some way of storing it during the second or so between heartbeats and then letting it go for the beat itself. However, development here has been very rapid and there will probably be practicable motors that can be implanted in the chest or abdomen by the mid-1970s.

But the motors have to be powered. Here, the first solution will probably be the rechargeable battery, or accumulator. This will be outside the chest, with a lead going through the skin to the motor, but there will be no need to trundle a barrow around or rush to a power plug every hour or two to get oneself recharged. It has been estimated that most artificial hearts will need between 30 and 50 watts of power to drive them (though the actual power needed to drive the blood is only about 1·5 watts while resting and 7 watts during moderate exercise). A pack of silver-zinc batteries weighing only six pounds could now provide 30 watts for 12 hours before they would have to be recharged, also for 12 hours, and with this kind of use would last a year before they were worn out. One of double the weight could of course take one through a day and night, while there would always be a stock of charged batteries at home for late nights.

So semi-portable, mechanical hearts might not be all that far off. However, many experts suggest that they will not be widely used because (other possible problems apart) the permanent leads through the chest will demand constant care and medical supervision. As with the kidney-machine patient there is a high risk of infection, but there is the more extreme hazard that the wires might break (as pacemaker wires often do), leaving the wearer about four minutes to die.

The first widespread use of mechanical hearts will probably begin only when those leads can be eliminated. At present, it looks as though this will probably not be done by implanting the power supply but by transmitting the energy from an external power pack through the closed chest by induction coils. This is already feasible: experiments have shown that at least 50 watts can be transmitted in this way without any medical problems. This 'radio link' power supply could be used in two ways. In one it would be the means of recharging *implanted* batteries, when they are possible. However, battery development is progressing very fast. For example, if the silver–zinc batteries already mentioned could be made to reach their theoretical limit they would give nearly four times the power for each pound weight, while there are silver-cadmium batteries – now very expensive and only in their early development stages – that could do much better even than this. The other way this radio link might be used is to pass current from a battery pack outside the body to an electric motor of some kind inside. So far the right kinds of motor systems have not been developed much, and when they are there will be still further problems to be solved. Most seriously, the coils that transmit and receive power – one set outside the chest, the other implanted – must stay exactly lined up with each other. If they slipped there would be a big power loss and, to say the least, the wearer might suffer. There is also the problem that they may be seriously upset by metal objects near by: lifts and cars could be a dangerous hazard.

Finally, there is the transition to the totally implanted heart and power supplies. For this there seem to be only two likely power systems at present. The first and nearest to readiness – some guess that it will be practicable well before 1975 – is the steam or hot-air engine, getting its heat from a radioactive isotope – a mini-scale nuclear power plant inside the chest.

If this sounds bizarre, the fact is that these devices have already been implanted in animals and have been scaled down to a size and weight not all that far short of acceptability for humans. For example, one design by the Thermo Electron Corporation (U.S.A.) is entirely contained in a three-inch diameter, four-inch long, four-pound cylinder. Probably this would be implanted in the abdomen and anchored to the bones of the pelvis, while a pipeline filled with gas or liquid would carry the pulses of pressure from the steam engine to drive the heart pump itself, up in the chest.

Oddly enough, the most serious problems for isotope-powered hearts are not to do with heat and radiation. There were no problems of this sort in the four patients (two French, two British) who in April to July 1970 became the first wearers of implanted atomic-powered heart pacemakers (though the power requirements of only 200 microwatts are somewhat different from those of a heart). Though a wafer of isotope may be as hot as 400°F and be throwing out potentially damaging high-speed particles, heat and radiation shielding seem to be manageable problems,* though of course there will have to be long-term studies of possible genetic and cancer damage. One of the biggest problems comes from the simple fact that all isotopes give out

* For the technically minded it should be said that no one is contemplating using isotopes that emit gamma radiation, as the screening needed to stop it would be prohibitively heavy. The choice is between alpha-particle and beta-particle emitters. While both kinds of radiation are easy to stop with very thin shielding, they create secondary problems that are not so easy. Alpha-emitters produce helium gas: how can this be vented into the body without sucking back some body fluids into the hot isotope chamber? They also sometimes undergo spontaneous nuclear fission, producing unstoppable neutrons that could cause cataracts in the eyes. With beta-emitters the main problem is that they produce secondary X-rays when stopped in the shielding, so the shields have to be made extra thick to stop the X-rays.

less and less heat with time as the radio-active atoms are 'used up'. So if they are to give enough power at the end of their useful life they have to give too much heat at the start when they are first implanted. Handling this initial heat overload and getting constant power from a slowly dying 'fire' could be difficult. But the overwhelming problem looks as if it will be the scarcity and cost of the isotopes. The most suitable isotope seems to be Plutonium-238, largely because its activity dies to a half of the original only every ten years, which means that if the engineers can handle twice too much power at the start the power capsule could be left in the chest for a decade before it needed to be renewed. But at the present time Plutonium-238 costs $1,000 a watt of power, or $30,000 to fuel a patient with a 30-watt heart. If this were leased by the government, including surgical fees for replacement, fuel reprocessing costs after ten years, interest and so on, the annual fuel bill for a patient would be around $2,000. And this is one of the cheapest fuels. Equally serious, it has been estimated that in 1975, when the first mass demand for powered hearts might come, the total production of suitable isotopes in the U.S.A. for all purposes will meet only about one-hundredth of the demand for hearts alone – if everyone in need got one.

The second possible implanted power source is a longer-term hope. It is to make a biological fuel cell that would draw its energy from the body: that is, from food. Like batteries, fuel cells convert chemical energy into electricity, but with the fundamental difference that the chemical fuels are fed to them continuously. A fuel cell for the heart would probably use glucose and oxygen as its fuels, taking them from the bloodstream, letting them react together on artificial membranes in the cell to produce a current that is picked up by electrodes, and then returning the non-toxic wastes back to the blood. It is a beautiful idea, but unfortunately at the moment no one has managed to make a reasonable-size biological cell produce more than about half a watt of power. Still, though it may be a long shot, the advantages are so great that there is bound to be intensive research into it.

So there, in broad outline, is how the development of the ulti-mate implanted artificial heart might proceed. But before we look at its implications, we ought to let the sceptics have their say. Some specialists believe that, however much ingenuity goes into them, implanted hearts will be physiologically impossible, or at best highly impractical.* One of their chief objections is over the matter of weight and size. No tissue can stand a heavy weight for long, least of all bone, which deforms under the slightest sustained pressure (which is why dental braces can straighten teeth). So it looks, the critics claim, as though any hope of trying to anchor heavy compact machinery to the skeleton – like that Thermo Electron isotope package, which is eight times heavier than the same volume of tissue – is ruled out. Instead, the engineers will have to make their machines no denser than tissue by encasing them in large cans of air. But this of course will make them big, perhaps unacceptably so. Even if the machines could be packed inside the skin, artificial-heart wearers will be recognizable by their huge stomachs.†

Another suggested limit is from heat. The normal heart generates 15 to 20 watts of heat, which is dissipated through the bloodstream. If an artificial heart can keep down to this, well and good. But to provide 7 watts of pumping power to the blood during normal exercise, they would have to be between 30 and 50 per cent efficient to do so. For small machines this is a very tall order. Recognizing this, the specifications for artificial hearts say that they should not generate more than 50 watts if possible and that to carry this heat away the blood temperature must not be allowed to rise above 105·6°F. Unless the engineers pull off a near-miracle, heart wearers may also be recognizable by their high fever!

To the extent that these and all the other problems are not solved satisfactorily, the widespread use of mechanical hearts will

* For example, see Donald Longmore's article 'Implants or Transplants', *Science Journal* (4 February 1968), pp. 78–83.

† In fact the agreed specifications for artificial hearts suggest that the volume in the chest for the heart pump should not exceed 1,000 ccs (1¾ pints) and for the power supplies in the abdomen, 1,000 ccs to 2,000 ccs, or up to 3½ pints. These figures are based on fairly optimistic assumptions that the weights can be fined down to 0·75 and 1 kilogram (1·65 lbs and 2·2 lbs) respectively.

be delayed. On the other hand, surgeons and doomed patients being what they are, many machine hearts will be inserted prematurely, before they give any reasonable hope of extending life for very long or with fair rehabilitation. While this will resurrect all the storms about heroic surgery that greeted the first heart transplants, it will also raise in new forms the 'failure' problems we met with kidney machines. Yet if we could somehow avoid these problems by jumping straight into 'good' mechanical hearts, our troubles would perhaps be even worse. For the 'success' problems of the artificial heart are formidable.

The overwhelming one, of course, is over supply and demand. Figures for those who might need artificial hearts of various kinds in 1970 have been carefully estimated for the U.S.A. – and since they are relevant to heart transplants they are worth looking at in some detail.

Out of the 800,000 Americans who will die from some form of heart trouble in the year, and the 2,035,000 who will be living with a seriously disabling heart defect, the bulk of the candidates for replacement hearts will come from people with coronary heart disease, that is, the heart-attack victim. The way this 'catastrophe' population could become potential candidates for artificial hearts is shown in Figure 9.3. From all the heart-attack victims who do not die before they can be helped, or who do not recover on their own, there would be 680,000 potential candidates for temporary support devices and 400,000 for permanent total heart replacements.

On top of this there are the victims of other kinds of heart disease who gradually degenerate rather than collapse suddenly. Here it is estimated that 141,000 will die in the year, all of them potential candidates for permanent rather than temporary support hearts. So for permanent hearts, whether half-heart or whole-heart machines (or transplants) we have 400,000 plus 141,000, giving 541,000 who face heart replacement or death during the year.

This is a ludicrously large figure. But, as with the kidney machine, it can be pared down by excluding the savable who are not 'worth' saving because of other medical complications.

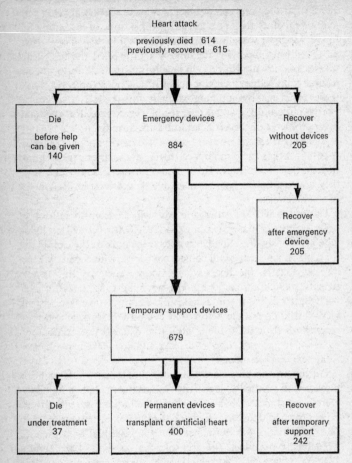

FIGURE 9.3 Numbers in thousands

**Heart attack**

previously died 614
previously recovered 615

**Die**

before help
can be given
140

**Emergency devices**

884

**Recover**

without devices
205

**Recover**

after emergency
device
205

**Temporary support devices**

679

**Die**

under treatment
37

**Permanent devices**

transplant or artificial heart
400

**Recover**

after temporary
support
242

Having done this, the numbers of *suitable* candidates are as follows:

TABLE 9.1

| | 'Catastrophe' Patients | | 'Decline' Patients | |
| | Potential | *Suitable* | Potential | *Suitable* |
| Age | Candidates | Candidates | Candidates | Candidates |
|---|---|---|---|---|
| 35 | 1,000 | 500 | 3,000 | 500 |
| 35–44 | 10,000 | 4,800 | 5,000 | 1,800 |
| 45–54 | 33,000 | 13,200 | 10,000 | 3,000 |
| 55–64 | 68,000 | 19,100 | 22,000 | 5,700 |
| 65–74 | 117,000 | 20,700 | 35,000 | 7,100 |
| Over 75 | 171,000 | 4,500 | 66,000 | 5,500 |
| | 400,000 | 62,800 | 141,000 | 23,600 |

On top of this there are those 2,035,000 cardiac cripples surviving in the population. It is impossible to guess how many of them would benefit from heart replacement, or want it, until it is known how well artificial hearts (or transplants) work and what life with them is like. But they are not necessarily bottom-priority candidates: many of the first heart transplants have been from this group. If one assumes that mechanical hearts do work well it is a reasonable guess that half who suffer a marked restriction in activity and all who are totally limited will want replacements. This gives a total pool of 461,000 potential candidates and, if this was cleared over ten years, a demand for hearts of about 46,000 a year for the next decade.

In other words, there are 86,400 suitable 'dead without' candidates, and 46,000 disabled candidates, totalling 132,400 implants a year. In addition, 884,000 patients would need emergency devices of some kind and 679,000 some kind of temporary supporting device, including temporary artificial half-hearts.

Could all these people get new hearts? A short answer is 'yes'. If society was determined enough that they should and there was a crash programme starting almost now to build the factories for making the hearts, to build the units for implanting them, and to train the men to implant them, it could be done.

Apart from development costs (put at between $60 million and

$200 million) and the even greater capital investment for training the teams and building the units, etc., most estimates are that the direct cost of an artificial heart implant would be $10,000 a patient. (Isotope hearts would cost roughly five times this, but with a programme to build nuclear reactors to produce Plutonium-238 these costs could drop substantially.) Once the production lines were going full blast, not much of this cost would be for the hardware itself, though at present prototype hearts cost about as much as cars, from Minis to Rolls-Royces, depending on their sophistication. Most of it is for the operation itself and post-operative care, while of course, unlike the kidney machine costs, these costs are once-and-for-all costs and will not accumulate from year to year if the hearts work well.

For all suitable patients that $10,000 a time adds up to about $1,500 million a year. Though this sounds astronomical it is only 0·15 per cent of the U.S. Gross National Product and 3 per cent of total health spending (in 1970). For Britain the equivalent cost would be roughly £150 million, well under one tenth of the Health Service budget, to save 33,000 deaths a year, or one-twentieth of all deaths. Is this too much, or not?

What about manpower? It has been estimated that to handle all the 132,400 implants in the U.S.A. each year there would have to be 650 surgical teams (cardiac surgeon, two assistant surgeons, cardiologist, biomedical engineer, anaesthetist, plus nurses, etc.) if they each did 200 implants a year, or 4 a week. This assumes perfect scheduling, with only one heart-attack patient clocking in each day, and that the teams would more or less do nothing else. It is more likely that there would have to be around 1,000 teams, or 3,000 skilled cardiac surgeons alone. It has been estimated that in the whole of the U.S.A. there are not more than 100 to 200 surgeons skilled enough at present to implant a mechanical heart. But again, if the build-up were gradual and the idea of heart implants attracted surgeons and society, it should be possible to train enough. By 1975 those 3,000 surgeons will, after all, represent only 3 per cent of all American surgeons – not a fantastic imbalance.

So it could be done. With a social investment roughly equivalent to setting up a new weapons system such as a sophisticated strike fighter, and making a hundred of them a year, artificial hearts could be provided for all. But we may not want to do it, not because of the cash and manpower it would use up, but because of all the lives – so miraculously saved – it would tie down. In other words, are the 'failure' problems going to be the worst of all?

The enthusiasts insist that before we ask this kind of question we ought to wait until there are people wearing hearts who can answer them. They are very fond of telling the story of the man with a loudly clicking plastic heart who was asked by a fellow passenger on the bus, 'Aren't you dreadfully embarrassed that we can all hear that thing?' and who replied, 'I'd be more embarrassed if you couldn't.' And, of course, they rightly point out that when you are facing certain death life of any kind seems worth living. But I think we ought to be aware that living with even a sophisticated fully implanted plastic heart will probably be socially and psychologically very disturbing. Indeed, there is not likely to be a single important aspect of life that will not be seriously altered by it.

For a start, the wearer's life hangs by a thread. Like the person with a pacemaker who dies when a wire breaks, a single malfunction can occur at any time with little forewarning and little that others can do to help. Many pacemaker wearers are deeply disturbed by this, though many adapt to it. No one knows whether it will be worse with a whole heart. Physical activity may be severely restricted – and not just every now and then, as it is with the kidney-machine patient in his off days, but continuously. This will demand difficult adjustments from the patient's family. With heavy or even moderately active work ruled out, social security support and vocational retraining will be essential in most cases. White-collar workers will be at an advantage here, as they are with the kidney machine. But to what extent will employers play along? Why should they hire a 'ticker' who may have to have a lift installed, might upset other employees, and whose family could sue them heavily for negligence if he died on the job? Families and employers will have to be brought into the rehabilitation process and possibly given state aid for equipment, etc. The heart-wearer's emotional responses may also be severely restricted: the control of anger, fear, anxiety, excitement, shocks – even with the aid of drugs – may be one of the most difficult life-adjustments to make. And if he has to adjust by living with only bland psychological stimuli, what kind of man will he become after a few years? Like any amputee, he may feel phantom pain from his old heart – but unlike the normal amputee he will not be able to see his lost limb or grasp the stump to help to control the pain. Worse, his heart will probably make a noise, give off heat, and visibly thump about. How will the patient, his family, his friends and colleagues and strangers accept this? With the symbolic significance of the heart in our culture, he might feel and be made to feel more different than most other kinds of handicapped person. Like so many handicapped people, he might have to withdraw into a small, intimate group of friends and relatives, or select friends only from others with plastic hearts. 'Ticker clubs' will spring up – possibly, it has even been suggested, special colonies where implantees would provide themselves with the particular medical and other props they need. Would the undoubted survival value of the colonies offset the apartheid stigma attached to them (plastic-heart wearers are hardly going to be an elite group)? And yet the patient is alive; without that troublesome contraption in his chest he would not be.

Many people are horrified at the prospect of plastic hearts or any other complex mechanical organs implanted in the body that biotechnology may develop. They see them as a threat to the most basic ideas of what it means to be human, the doorway to a race of de-humanized Mechanical Men. At the other extreme there are those who say we shall soon get used to them, as we have got used to every other mechanical prop of civilization. Like the Australian aborigines who react with terror when they first see a man with false teeth but soon clamour for them, we shall quickly adjust our ideas about what 'natural' and 'unnatural' mean.

What I hope this chapter has shown is that the truth lies in between and will be far more difficult to meet. Because for generations to come our plug-in and stitch-in machines will be such pathetically inadequate copies of the real thing we can discount all those speculations that we shall use them for the sheer hell of it – to create super-athletes with double hearts, for example, or cyborgs with living brains but non-eating, non-excreting mechanized bodies for space journeys. Yet because they will be adequate to sustain life of a kind, it is flying in the face of human nature as well as medical ethics to suggest that they will not be developed and used, that many people will not want them very much indeed, and that we shall not be faced with all the problems of helping them cope with their 'half-lives', or refusing them because they have little chance of coping well or because we decide that our resources are better spent elsewhere. Other issues for which there is no space here, such as the impact successful spares will have on the age structure of the population or on the numbers who will be unemployable, will be no less perplexing. We should have started talking about these things years ago, and at a level accorded to any other major strategic decision, if we are to hope to manage them successfully.

Fortunately, though, we may have a chance of avoiding many of these awful problems if only we can learn to manage the other half of the spare-part business – the transplanting of natural organs.

# 10. TRANSPLANTS

The great hope of transplant surgery is that when it works it ought to work well because the spare parts it uses are designed to work well. Engineers may develop ingenious machines to replace failed organs, but for a very long time to come they are going to be cumbersome, inefficient and vastly inconvenient death-postponers rather than life-restorers. How much better to use natural parts that have a hundred million rather than ten years of development behind them; parts that are as complex and adaptable as the machine that houses them; and parts that are free.

And yet, of course, it is not as easy as that. Transplant surgery – whether it often fails, as now, or whether it almost always succeeds, as it might soon – raises some of the sharpest problems in the whole spiky jungle of modern medical dilemmas, many of them arousing deep anxieties and emotions. It is not easy to face them calmly, but as we launch into a discussion of a major part of our biomedical future we have to sort out what the real and what the imaginary anxieties are, so that we can set a course that will bring us maximum benefits with the least costs.

So far we have not managed this discussion well. The most striking thing about the whole transplant business until now has been the way perhaps nine-tenths of the public discussion about it has been based on a few score transplants of a single organ, the heart. One might as well debate what should be done about world transport after seeing a multiple pile-up on a motorway. The first thing that urgently needs doing, therefore, is to set transplanting in a broad time perspective, to give it the crucial flavour of change – and rapid change at that. And when we do this we shall find several striking things emerging; not least the way technical improvements have had a profound effect on the deepest moral problems, making some of them more relevant, some of them less.

One could start a long way back. Medea is supposed to have arranged a transplant – of blood – between Jason and his father; while skin grafting from one part of the body to another was carried out in India as far back as 600 B.C. Even with the major, compact organs one still has to go back over sixty years to the first human experiments. Before the First World War several European surgeons had grafted kidneys into humans, using rabbit, goat, pig and primate donors. The American, Charles Guthrie, had even achieved the Everest of transplantation surgery by producing a two-headed dog (though it only lived a day). As for attitudes, the whole shutter-clicking, headline-making, mud-slinging, glorifying ballyhoo of the first heart transplants, with nearly all the deep emotions it generated, had already been ex-perienced and quickly forgotten by all but a few religious sects with blood transfusions back in the 1920s.

The best starting-point is the first kidney transplant of modern times, in the early 1950s. There are several good reasons why the kidney should be the archetype transplant organ and why so far about 90 per cent of all grafting of major organs has been and still is concerned with it. Surgically, kidney grafts are easy; there is usually only one vein, one artery and the ureter to join up. A new kidney seems to function perfectly well without a nerve supply. We have two kidneys, so that living donors, who give the best chance of success, can be used. Above all, there is the kidney machine to get the recipient as fit as possible before the transplant, to let him wait in a queue for a suitable donor to appear, and to help him should his new kidney fail.

The first modern kidney transplant was performed in April 1951 by Dr David Hume at the Peter Bent Brigham Hospital, Boston. This was frankly experimental surgery. There were no anti-rejection drugs and very primitive ideas about matching donors and recipients. So for ethical reasons the patient chosen was in the final, inevitably fatal stages of acute kidney failure; and the donor was a recently dead person.

For the next four years these two ethical criteria dominated transplanting. In that time Dr Hume and Dr Joseph Murray carried out ten human kidney grafts, all with cadaver organs and patients on the verge of death. Six of the ten kidneys failed to function after the operation and the patients quickly died; with the four that did work two patients survived about a month and one for six months. Note that this is a worse record than for the first ten heart transplants (see Table 10.1). The most important thing to emerge from these experiments was the crucial importance of ischaemic damage – the deterioration of an organ when it is not connected to a warm host body. During the series, techniques for cooling and perfusing the kidney with fluids during the transfer were developed. These gave better survival figures, lowered the last major *surgical* barrier to success and became standard practice for all transplants.

The next step in transplanting involved a major ethical leap. Transplanters knew that the ideal way of overcoming the graft rejection barrier was to use an identical-twin donor. Because he and the recipient would have identical genes, the recipient's immunity system would not recognize the graft as a foreign body and would make no attempt to reject it. But one of the fundamental canons of medical ethics is *primum non nocere*: first of all, do no harm. To use a living donor would break this rule: for the sake of one patient it would mean the deliberate mutilation of a healthy person. How should this be balanced against the overwhelmingly better chances from using a twin rather than a cadaver?

The balance was first tipped to favour the dying patient on 23 December 1954, when Dr Murray and Dr John Merrill did the first ever living-donor transplant at the Peter Bent Brigham Hospital. It was also the first wholly successful transplant. The patient, Richard Herrick, in the terminal stages of kidney disease at twenty-three, received a kidney from his twin Ronald, went on to a virtually complete recovery, married his nurse, became a father and lived for eight years before he died of a heart attack resulting from his original kidney disease.

Spurred by this success, kidney transplanting entered its 'live

**TABLE 10.1**  Heart transplants: the first 20 or first 6 months

| | | | |
|---|---|---|---|
| Louis Washkansky (55) | Cape Town, South Africa | 3 December 1967 | 18 days |
| Boy (19 days) | New York, U.S.A. | 6 December 1967 | $6\frac{1}{2}$ hours |
| Philip Blaiberg (58) | Cape Town, South Africa | 2 January 1968 | $84\frac{1}{2}$ weeks |
| Mike Kasperak (54) | Stanford, U.S.A. | 6 January 1968 | 15 days |
| Louis Bloch (58) | New York, U.S.A. | 9 January 1968 | 10 hours |
| Bodan Chittan (27) | Bombay, India | 17 February 1968 | 3 hours |
| Clovis Roblain (66) | Paris, France | 27 April 1968 | 2 days |
| Joseph Rizor (40) | Stanford, U.S.A. | 2 May 1968 | 3 days |
| Everett Thomas (47) | Houston, U.S.A. | 3 May 1968 | 29 weeks (died 2 days after 2nd transplant) |
| Frederick West (45) | London, U.K. | 3 May 1968 | $6\frac{1}{2}$ weeks |
| James Cobb (48) | Houston, U.S.A. | 5 May 1968 | 3 days |
| John Stuckwish (62) | Houston, U.S.A. | 7 May 1968 | 7 days |
| Elie Reynes (65) | Montpellier, France | 8 May 1968 | 2 days |
| Fr Damien Boulogne (45) | Paris, France | 12 May 1968 | 76 weeks |
| Louis Fierro | Houston, U.S.A. | 22 May 1968 | 20 weeks |
| Joseph Klett (54) | Virginia, U.S.A. | 25 May 1968 | 7 days |
| Joso de Cunha (23) | Sao Paolo, Brazil | 26 May 1968 | 4 weeks |
| Albert Murphy (59) | Montreal, Canada | 31 May 1968 | 1 day |
| Antonio Serrano (54) | Buenos Aires, Argentina | 31 May 1968 | 4 days |
| Ronald Smith (38) | New York, U.S.A. | 1 June 1968 | 1 hour |

donor era' – a stage we are still in but which many transplanters are struggling to get us out of. We shall see why, and how, later. The crucial point for now is that for the next seven years all transplant results (apart from identical twins) were appallingly bad, but despite violent attacks the surgeons went on transplanting. The world figures for transplants performed up to September 1962 show starkly what was happening (see Table 10.2).

TABLE 10.2   World kidney transplant survivals at September 1963

|  | Patient still alive/Patients transplanted (and as %) | | | |
|  | 1 year after | | 2 years after | |
| Kidney from: | | | | |
| Identical twin | 21/25 | (84%) | 18/22 | (81%) |
| Close relative | 10/94 | (11%) | 3/90 | (3%) |
| Cadaver (unrelated) | 1/59 | (2%) | 0/58 | (0%) |

Quoted by Professor M. F. A. Woodruff, 'Ethical Problems in Organ Transplantation', *British Medical Journal*, 1 (1964), 1457–60.

Then just as the critics were beginning to get extremely acid, transplanting broke into a growth curve of success. The reason was the discovery of the first powerful anti-rejection drug – azathioprine (Imuran) – first used on a transplant patient in March 1961.

Most people now know in outline what graft rejection is about. If one introduces almost any fragment of living material from another organism into the body the new host will recognize it as 'non-self' and will try to destroy it. This immune response is a valuable defence strategy against invasion by infectious organisms, dirt and even cancer – abnormal mutant cells that become genetically non-self inside the body. In more detail it works like this. Throughout the body there is a network of lymphoid tissue. When foreign material reaches it through the bloodstream it can recognize the intruder by the particular types of proteins – called antigens – that it contains. The lymphoid tissue reacts to these antigens by producing antibodies to neutralize them. Other parts of the defence system then destroy the invader, while the immunity system 'remembers' the attack so that the next time the same set of antigens gets into the body it swings into action more rapidly and powerfully. This is the basis of all long-term immunity against disease.

With an organ graft the strategy is rather different. Usually an implanted organ produces only a trickle of antigens and these stay close to the graft. So the response is far more local. The body detects the foreign antigens using the outriders of the immune system – mainly cells called lymphocytes. These circulate in the bloodstream, visit the graft, are activated there by the foreign antigens and then as they pass the local out-stations of the lymphoid system (the lymph nodes) start up a local counter-attack. The upshot is that the cells of the graft are attacked and its blood supply is choked off so that it 'dies'.

One way round this transplant barrier is to knock out the whole immunity system with anti-rejection or immunosuppressive drugs. Before 1961 the only available drugs were the cortisone group, especially one called Prednisone. But they had little effect and in the doses used to try to stave off rejection produced severe side-effects. Imuran added a powerful second attack force. Though it too has side-effects – particularly anaemia and muscular wasting – they are different from those of the cortisones. So in combination the two drugs add up to a powerful rejection-preventer that can be used in doses low enough to minimize side-effects. One side-effect that cannot be prevented, though, is that by knocking out the whole immunity system these immunosuppressive drugs leave the patient wide open to infection. Even a simple sore throat can burst into a generalized lethal infection – and has done – unless antibiotics are given quickly.

These drugs are still the main weapons against rejection, and much of the art of transplanting is striking the delicate balance between giving enough of these drugs to prevent rejection and not giving so much that infection or the drugs themselves kill the patient. The really critical time for this balancing act is the first four months after the graft, when nearly all severe rejection crises occur. The usual strategy is to give as low doses as one dares and to monitor the patient very carefully for early signs of rejection.

If these appear, the drugs are then stepped up, perhaps to massive doses during a severe rejection episode.

This can be a harrowing time for the patient. He will almost certainly be in hospital, subjected to innumerable tests, with the nearly constant threat of failure and death (though this is no new threat to him). But if all goes well, the patient leaves hospital with his drug dosage optimized and his new organ working; daily out-patient visits become weekly, then monthly, and he can nearly always return to a normal life – in sharp contrast to the man on a machine. What happens after that, over the next years, varies enormously. Some patients will have to go on with fairly high doses of anti-rejection drugs and may have serious medical troubles from their side-effects. Sometimes organs will be rejected despite these doses. Sometimes, as with the Herrick 'first', the patient will be killed by the after-effects of the disease which brought him under the transplanter's knife. But in many cases none of these things happen and – perhaps because of good donor-recipient matching (whether by luck or design) or perhaps because the recipient's body gradually adapts to his foreign graft – drugs can be cut to almost unnoticeable levels with no risk of rejection. For these people a transplant is in the fullest sense the gift of a new, full life.

To return to Imuran, its arrival improved transplant results dramatically. Within a few years the 3 per cent and 0 per cent two-year survival figures for related and cadaver donors had leaped to 60 per cent and 40 per cent in the leading centres.* This was still far from 'successful', but it was good enough to have several major consequences. For one thing, the number of

* The latest available figures show that *nearly all* centres now do about 15 per cent better than this. With the 736 transplants performed in 1966 and 1967 by the 99 centres reporting to the central registry, the results were: sibling donors, 75 and 67 per cent survived one and two years (pre-1966 averages were 60 and 50 per cent); cadaver donors, 45 and 38 per cent survived one and two years (pre-1966 figures were 25 and 20 per cent). Note that survival refers to the new kidneys: with kidney machines and retransplanting average patient survival is considerably better. There was hardly any difference between results from experienced and inexperienced centres. *Transplantation*, 6 (1968), pp. 944–56.

transplants started to grow explosively – giving a hint of what to expect in the next few years with heart, liver and possibly lung transplants:

TABLE 10.3

| Period | Number of transplants | Average number each year |
|---|---|---|
| Dec '54–Mar '63 | 153 | 19 |
| Apr '63–Mar '64 | 239 | 239 |
| Apr '64–Mar '65 | 280 | 280 |
| Jan '66–Dec '67 | 736 | 368 |

Second, there began a steady swing from live donors to the ethically less objectionable use of dead donors. Some centres even started refusing to use living donors, insisting that though dead donors might not give such good results, for ethical reasons transplantation had to learn how to break through into the 'dead donor' era. Third, with a perhaps premature faith in the powers of anti-rejection drugs but with their eyes on the distant future, some surgeons closed the circle and started using animal donors again. Fourth, the early experiments in transplanting human livers, lungs and hearts began.

We shall come back to these topics later. But before we do there are two crucial and recent technical advances to look at: the latest and perhaps ultimately effective anti-rejection agent, called anti-lymphocyte serum; and tissue typing.

Compared to the older anti-rejection drugs, anti-lymphocyte serum (A.L.S.) is a rapier rather than a block-buster. It attacks only the lymphocyte 'scouts' and so leaves the main immunity system intact. No one is quite sure yet how it works, but the two main theories are that it depletes the number of lymphocytes and/ or coats their surface so that they are immunologically blinded. The serum is prepared by injecting human lymphocyte cells into animals. The animals react by producing antibodies against the lymphocytes: in other words, they make an anti-human lymphocyte serum which can be extracted from the animal's blood, purified and injected into a human patient to keep his lymphocytes quiet.

Experience on human patients with the drug has not been large, and is anyway evolving rapidly, so it is too soon to assess what it might do for long-term survival rates. However, from the trials that have been done by the pioneers – notably Professor Woodruff in Edinburgh and Dr Starzl in Denver, Colorado – many transplant surgeons are now saying that with A.L.S., the standard drugs and good donor-matching 90 to 95 per cent of kidney patients should survive a year after their transplant and the vast majority of these should survive indefinitely. If this is so, kidney transplanting will be a better treatment than the kidney machine. Mortality will be lower and, with only a regular course of pills to take, the patient's life will be much less restricted. There is even the hope that A.L.S. will hoodwink the host body into *permanently* forgetting that the graft is there, so that even the pills can be put aside. A.L.S. also holds out the prospect that transplants of animal organs to humans might soon be routinely successful, thus breaking through the final transplantation barrier and opening the way to a limitless supply of donors.

Of course A.L.S. may not turn out to be the miracle answer, and even if it does it will probably take years to iron out the snags. The serious ones, at the moment, are that it is very difficult to prepare in sufficient quantities and with sufficient purity to avoid serious side-effects; and there is growing evidence that the use of A.L.S. (and other effective immunosuppressive drugs) makes the onset of cancer more likely. Because of these and other highly technical complications, A.L.S. was still considered to be highly experimental – at the clinical trial stage only – during the first half of 1968 when the early heart transplants were being made. Perhaps the survival rates would have been better if A.L.S. had been ready for use.

The aim of tissue typing is to lower the transplant barrier from the start by selecting a donor whose antigens match the recipient's. This is possible because although every individual has a unique, personal set of antigens, only a few 'strong' ones among them really matter. So just as blood can be transfused safely between strangers because it is technically easy to type people into the main ABO blood groups – that is people sharing the same strong blood antigens – tissue typing is concerned with finding quick, simple ways of matching donors and recipients for strong tissue antigens. It is now known that the human race is divided into about six to eight such groups (though, as with blood, further subdivisions are sure to be made), and that transplants between these groups are far less likely to be rejected than grafts across them. As with blood transfusions, one can range from rare 'perfect' matching through shades of 'good' compatibility to complete incompatibility.

The most promising method is the white blood cell (leucocyte) test. Developed during 1967, it depends on having banks of blood sera from a wide range of individuals who have antibodies against human tissue in their bloodstream. Samples of these sera are mixed with leucocytes from the person whose tissue is being tested. If the mixture 'clumps' there has been an immunological reaction: the antigens are dissimilar. If it remains clear the antigens are similar. It sounds cumbersome but there are now plastic kits available which allow a donor or recipient to be typed against the bank samples – and therefore against each other – in a few minutes.

Given further refinements of tissue typing and A.L.S., most experts predict that the rejection barrier will be broken – at least for human-to-human grafts. There should be no immunological reasons to prevent the grafting of kidneys, livers, lungs and hearts (and a good many minor organs) at will. In other words, in less than two decades from its scientific beginnings, transplantation will have rushed ahead of most of the moral and social problems deriving from its *failures*. And we shall be faced with the even more fearsome social and practical challenges deriving from its *success*.

Before we look at these let us make a quick survey of the other major organs apart from the kidney: namely, the lungs, liver, heart and brain. Apart from the obvious points that these are single, vital organs and can come only from dead (or perhaps dying) donors, and that there are as yet no supporting machines for them when a graft fails, what are their histories and what special problems do they raise?

*Livers*

Transplanting a liver is a formidable surgical feat and few have attempted it. In fact since Dr Thomas Starzl did the first in March 1963 his team at Denver, Colorado had (by November 1969) done twenty-five of the world total of fifty. But while Starzl's first seven patients all died within three weeks, six of his last eighteen patients survived more than a year. Other surgeons are showing similar improvements in their success rates.

The main trouble is that to graft in a liver involves a mass of delicate plumbing in some very inaccessible places, and it takes about six hours to do carefully. There is a risk of lethal blood clots on the one hand and massive haemorrhage on the other if anti-coagulants to prevent clots are maintained. Worst of all, the liver is the most susceptible of all organs to damage while outside the body. Twenty minutes without cooling or perfusing and it will deteriorate enough to considerably raise the chances of rejection; an hour and it will be so badly damaged that there is little hope that it or its new host will live.

On the brighter side, if it is not damaged the chance of success is high, because there is strong evidence that it is particularly resistant to rejection. Several pigs have survived for months with livers from unrelated donors without ever being given anti-rejection drugs. In fact some liver surgeons now think that they would have been much more successful with human liver grafts if they had used lower doses of anti-rejection drugs. Also, the liver resembles the kidney in that there are no problems about nerve supply either.

Apart from training a breed of highly specialized surgeons, therefore, the main problem with the liver is to prevent ischaemic damage. This will probably be solved either by chilling the dead donor's body until the surgeons are ready to graft it into the recipient, or taking out the liver soon after death and storing it in a refrigerated perfusion chamber. To do this successfully will not be easy. After it is done there is still the problem of cooling the liver while it is planted into the new host. Various ideas for refrigerated boxes that house the liver but not the vessels that have to be joined up are being developed.

When all this is done there is still a further problem. A chilled liver takes about two days to recover its full function after a transplant, and the recipient must usually be supported during this time. Ideally, as with the kidney, there ought also to be some way of enabling him to wait while in acute liver failure for a donor to die. Both factors point to the need for a 'liver machine'. Unfortunately, an artificial liver will probably never be possible: the liver is a fearfully complex biochemical factory making more than five thousand products, and the prospect of copying it mechanically is extremely remote. So instead an answer is being sought in hooking the patient's bloodstream up to animal livers – usually calves' or pigs' – to give temporary support for a few days. So far this has not been at all successful, and of course it can never become equivalent to long-term support by a kidney machine.

Because of these special problems, liver transplantation may be rather slow to develop on a large scale, though in Britain current plans are to do one roughly every six weeks.

## Lungs

The first human lung transplant was carried out in 1963 by Dr J. D. Hardy of the University of Mississippi Medical Center on a prisoner serving a life sentence for murder with chronic kidney disease and lung cancer. He died eighteen days later. By September 1968 the total had risen to ten, including Europe's first (at Edinburgh, on a boy who had swallowed weedkiller). All failed, no patient surviving three weeks. Since then there have been a few more attempts with slightly better results, though one (in November 1968 on a Belgian metalworker) kept the patient alive for ten months. Despite this slow start some transplant surgeons predict much better results soon. Others say that the lung may prove the most difficult major organ of all.

There are two major but related obstacles to successful lung grafts. The first is to do with nerve supply. Unless a part of the recipient's own lungs are left in, he will have little nervous control over breathing. He will be able to breathe only slowly and deeply, and his life, though possibly saved, will be severely restricted. But if a part is left in, one aggravates the second major problem: infection. It is very unusual for anyone to have sterile lungs. So the new lungs from the dead donor will almost certainly be infected, as will the recipient's remaining part-lung, or his trachea or bronchial stump. Apart from all this, a lung is virtually an external organ, so that even if a graft starts off sterile, with the standard anti-rejection drugs the recipient has little chance of escaping a lethal infection; two of the first three lung-graft patients died within a week of broncho-pneumonia. A.L.S. with its rapier action, coupled with antibiotics, could make a big difference here.

On the other hand, lungs suffer little ischaemic damage and are very easy to preserve and transport. This should greatly ease the logistics problem when, or if, lung transplants become commonplace. About all one needs to do is put the lung in a plastic bag of iced water, pump air into it with a small mechanical respirator, and perfuse the network of blood vessels surrounding the lungs through the stumps of the pulmonary vein and artery (where they have been cut off from the donor's heart). The whole apparatus can be little larger than a suitcase. Using a device of this kind a group of American doctors at the Imogene Bassett Hospital in New York stored the lungs of fourteen dogs for twenty hours and achieved four successful grafts when they transplanted them into unrelated dogs. Significantly, when the four survivors were examined eight weeks later the lungs were functioning well. Dr J. D. Hardy has done several hundred lung grafts on dogs and shown that grafted lungs usually work at near normal efficiency within a few weeks. During this time patients could be supported by respirators and oxygen-enriched air. Commenting on his work in November 1967, Dr Hardy – the pioneer of animal and human lung grafts – was optimistic about human prospects: 'Lung transplants are inherently feasible, and I feel that with continued improvement in immunosuppressive measures ... exploration in this field is well justified and should be rewarding.' It is difficult to be any more definite than that at the present time.

## Hearts

The first human heart transplant was done on 23 January 1964 by a team under Dr J. D. Hardy. The sixty-eight-year-old recipient was in a final state of cardiac shock, attached to a heart-lung machine, and had only an hour or so to live. Hardy had been hoping to graft a human heart into him and taken all the necessary surgical and legal steps. But the prospective donor, a young man dying of irreparable brain damage, still lingered. So on a vote (one abstention) the team decided to attempt the double first of grafting in the heart of a chimpanzee. They did so. The chimpanzee heart beat strongly for ninety minutes, then stopped, and the patient died.

The first well-publicized human heart graft and the first to use a human donor was performed (needless to say) by Dr Christian Barnard on 3 December 1967, with Louis Washkansky as the recipient, and a road-accident victim, Denise Darvall, as the donor. The floodgates were opened. Within hours the headlines were screaming and the world launched into a furious debate on heart-swopping based on one case, and on taking organs from the 'dead' sixteen years after it had first been done with kidneys. Within weeks Dr Barnard had met the Pope, kissed Sophia Loren, and been awarded an annual prize by South Africa for good public relations. Within months the number of heart transplants performed had started to climb as dramatically as the press coverage on each began to shrink. It took five months to the day to bring the total to 10, six months to bring it to 20, nine months to 40, and exactly a year to make a total of 95 grafts into 93 patients (two patients received a third heart).* But then the initial spurt slowed down. Nine months later, in early October

* Of the 95, over 50 were performed in the U.S.A., 12 in Canada, 4 in France, and 3 in South Africa. The remainder took place in eleven different countries – Argentina, Brazil, Chile, Czechoslovakia, England, India, Japan, Spain, Turkey, U.S.S.R., Venezuela.

1969, the total had risen by only 47 to 142. Almost a year after this, in September 1970, the total had risen to a little over 170. Disappointed themselves, under fierce attack from others, heart transplanters settled down to steadier programmes of human – and animal – experiments.*

How successful was this first burst into heart grafting? Table 10.1 shows the results of the first twenty: two lives prolonged well over a year, two about six months – a far better record than the first decade of grafting kidneys from dead donors, now standard clinical practice (see Table 10.2). By April 1969, when the total stood at 122 transplants, the record seemed to be improving (despite outcries to the contrary). Of the 33 patients still living, one (Blaiberg) was alive more than a year after operation, 13 were alive between six and twelve months after, 13 between three and six months, and 6 less than three months after. Many of these might live for long periods, while many of the dead patients had survived for several months rather than mere weeks, as earlier.

But then a wave of deaths worsened the record and the doubts hardened. Just over a year later, in September 1970, only 22 patients still survived out of a total of a little more than 170 transplant operations. Though several patients had lived or were still alive two years after their transplant, the one-year survival rate was only 10 per cent: very much lower than the comparable figure of 45 per cent for kidney transplants from dead donors. However, there was one notable exception to this record. Professor Norman Shumway, of Stanford University Medical Center, the true pioneer of human heart transplantation because of his

---

*The first animal heart transplant was carried out by Alexis Carrel in 1905, when he grafted a puppy heart into the neck of a dog (whose own heart was not touched). The first true heart replacements were done by W. B. Neptune in 1953 on three dogs. The longest survival was three hours. In 1960 R. R. Lower and N. E. Shumway started a major programme of dog heart transplants and worked out most of the techniques now used on humans. In the next five to six years many others followed their lead. So human heart transplanting was hardly a leap into the dark. On the other hand, up to the time of the Barnard 'first', nearly every animal getting a heart transplant died within a week of surgery. The occasional six-month and one-year survivals were very rare exceptions.

immense experience working out the techniques with animals, had by the same date scored a one-year survival rate of 35 per cent – three times better than the world average. Of his 25 heart-transplant recipients, eight were still alive, one more than 22 months after the operation.

The heart may be an emotionally and ethically loaded organ to transplant but technically it presents no unique problems for a skilled team. A heart graft is an easier surgical feat than many almost routine multiple heart-valve operations, and is far easier than a liver graft. Ischaemic damage is a problem – heart tissue is very vulnerable to lack of oxygen – but deterioration is not serious unless the heart is disconnected for about thirty minutes (twice the time for the liver, about the same as the kidney). However, two skilled teams can remove a donor heart and finish stitching it into a recipient inside forty minutes, while for most of this time the new heart can be fed with blood. Preserving the vital nerve supply to the heart is no great problem. The heart's natural pace-maker that keeps it beating automatically and speeds the rate up to pump more blood when needed is mostly in the upper right-hand part of the heart, called the right atrium. The right atrium of the recipient is left in place, and the new heart, with its right atrium removed, is stitched on to it. Thus the recipient keeps his own heart's nerve supply to drive the new heart.

Looking ahead, there are good prospects that, like kidneys and lungs, hearts will be able to be banked for some time. Using simple cooling and perfusion chambers, dog hearts have been success-fully transplanted after being stored for six to eight hours. With greater refrigeration in high-pressure oxygen chambers there have been excellent results after storage for twelve hours and 'quite good' results after twenty-four. If this kind of thing can be done with human hearts, it would no longer be necessary for the recipient to be surgically prepared and waiting in the theatre for the donor to die. This would lessen the vulture-like aspects of the death watch. It would also allow hearts to be transported and thus widen the pool of (matched) patients for every donor heart. Even further ahead (as we saw in the last chapter), there is a fair chance that a long-term heart-lung machine will one day be possible. While this would make heart transplants ethically more acceptable it might also reduce the demand for them.

The first years of human heart transplants also raised one very old but peculiarly vicious problem of medical ethics. It emerged about five months after the Barnard 'first' when the score was fifteen, with ten deaths and five patients still living. None of the recipients, living or dead, seemed to have shown any serious signs of rejecting their hearts. Then why did the ten die? With Louis Washkansky certainly, and probably with some others, infection as a result of anti-rejection drugs was the cause. But at least five of the ten appeared to have died because the lungs, liver, kidneys or brain, severely damaged as a result of the patient's long-standing heart disease, failed under the strain of having a fresh, strong heart to cope with, or were already beyond hope of recovery.

Many patients who were ill enough to qualify for a heart transplant may therefore have been too far gone to survive one. And yet they could only qualify because they were far gone: when embarking on a dangerous new line of frontier surgery, surgeons rightly insist that they should only take last-ditch patients who have little to lose if the operation fails. Younger, fitter patients who could benefit from the operation but would still live several years if they did not have it, are not taken on until the operation has been well tried and shown to work reasonably well.

Medical ethics therefore places the pioneer surgeon in a deep trap. He must develop a new operation on the worst possible type of patient and then make an agonizing judgement that from this experience he is ready to jump out of the trap to start trying his hand on more promising patients who have more to lose if he fails. With heart transplants, it was Professor Shumway who first took this brave step by announcing in 1969: 'we aren't going to do any more transplants in dying patients whose lungs, kidneys and other organs have accommodated themselves to a failing heart'. His unique success rates suggest that he was right to be brave.

## Heart plus lungs

Some heart surgeons (notably Donald Longmore) foresee heart-plus-lung grafts superseding heart-only grafts. The surgery is much simpler: instead of having to make eight joins there are only three (the trachea, or windpipe; the aorta, or the artery that takes blood from the heart; and the atrium nerve centre at the back of the heart, which also contains the two veins returning blood to the heart). Also, the heart and lungs are beautifully adapted to each other. One of the main lethal hazards of heart-only grafts is that a powerful fresh heart has to pump blood through a lung circulation with too narrow or too wide a bore; another is that if the recipient's original heart is damaged his lungs probably are too. Replacing the complete system avoids both these dangers, but we shall have to wait to see what other dangers (like lung infection) it introduces.

## Brains and heads

Brain and head transplants are impossible. First, there are no possible donors, dead or alive: the brain of a dead donor is, by definition of his death, dead; the brain of a live donor is not on the market. Therefore the only conceivable operation is a reverse whole-body graft: a patient with a sound brain and a smashed body receives a new body for his brain to run: nice science fiction but technically impossible. There are ten million nerves running from the brain down the neck (and millions more joining the brain to the sensing organs, if one is thinking of dissecting out a brain for transplantation into a head). Joining these would be an inconceivable task, and even assuming it could be done, those severed nerves would take months to regain their function, if indeed they ever did so.

Now let us look at the tangle of moral, economic, legal and social problems that transplanting presents us with, and will present us with in the future.

The first thing that must be said about them is that, apart from the problems of donors, none of them are new in principle. As Professor Keith Reemtsma, the American surgeon who has pioneered animal-to-human transplants, has said:*

I do not doubt that the first cave man who trepanned a skull was assailed for trying an unproved operation, for proceeding without conclusive animal experiments, for forsaking the medical regimen of powdered owl feathers, for getting too much publicity, for siphoning off public funds for his work and denying support to the tiger-tooth necklace project, for interfering with the mysterious plans of the Great Spirit, and, of course, for prolonging the lives of unfit individuals and thereby placing in jeopardy the future of the Cro-Magnon race.

Nevertheless, while people have been arguing about these kinds of problem since the caveman days, transplanting forces us to argue them afresh. So there is no justification for the blandly complacent attitude that they are best left to the experts to solve.

### Are transplants over-heroic surgery?

Many people believe they are. They thought kidney transplants were in the old days and they think heart transplants are now. They believe it is morally wrong to put a dying patient through the trauma of a major operation and a difficult recovery when there is so often so little point. Instead the patient should be allowed to 'die in dignity'.

This kind of attack has always been made on new, high-risk treatments. Nearly always, though, the attackers ignore several crucial points. For one thing, patients do not always 'die in dignity' if they are not operated on: death from kidney failure or untreatable heart disease is often far more lingering and painful than death from a failed transplant. For another, 'experimental' surgery is almost invariably performed on patients who are bound to die anyway, and soon. Thus the operation offers hope, however minimal. As long as the hopes are not exaggerated for the patient or his family, what harm is there in that? Critics forget that one cannot make general charges that this operation or that

* *Annals of Internal Medicine*, 61 (1964), p. 357.

is aggressively heroic. As we saw in the previous chapter, one can only make them for specific, individual cases on the basis of whether or not the patient gives his full, informed consent.

Having said that, some important general questions should be raised. One is whether fully informed consent is ever truly possible with high-technology medicine. As medicine produces more extraordinary death-postponing techniques it becomes more difficult for patients to understand all the subtle technical implications of alternative treatments. It becomes easier for patients and doctors to give up trying to communicate and to fall back on the old, old notion that 'Doctor knows best.' But the increasing 'power of medicine', by presenting more and more complicated alternatives, widens the gap between the experts and the ignorant patient, the power of the mighty over the weak. And though most doctors try to bridge this gap as best they can, and try to use this power benevolently, it is inevitable that in the end the power and decision is theirs. Consider, for example, this quotation by a leading American transplant surgeon, Dr Francis D. Moore:

> It is not enough to tell the patient that 'there is no other hope'. If he is in full possession of his faculties, he should be given a clear picture of the hazards involved and allowed to join in the discussion. Yet under no circumstances should the final decision be left in the hands of the patient; he has not the education, the background nor dispassionate view necessary to make a decision in his own best self-interest.

And how dispassionate can the surgeon be? How much can he ignore *his* self-interests? As long as surgeons are human they will be driven by ambition to do 'firsts', will want to practise on humans so that untried operations can become routine, and push their own favourite operation. In the atmosphere of a go-ahead transplant unit is it possible for the surgeons not to be over-optimistic to their patients, not to favour transplanting over possible alternatives?

A second general point is whether in our culture a dying patient can make a free choice about what should be done to him. We are all so dedicated to the idea of not dying that when we are at death's door it must be nearly impossible to judge whether life after a successful operation would be worth living. To give an extreme example, there is an operation called a forequarter amputation in which patients with extensive cancer of the lower half of the body are cut in half at the waist. This may prolong their lives by a few months, or even years, but it is a terribly crippled life. In a recent case, when asked why he had allowed himself to be so terribly mutilated, the patient replied simply, 'I wanted to live.' One sees his point, and in a sense his answer stifles all criticism. But does it not also suggest that now the patient is stuck with the results of the operation he could give no other answer? And that his doctors – also dedicated to death-postponing – could not see beyond the immediate, narrow, medical crisis?

There are no easy solutions to these dilemmas short of a radical change of attitudes – the change that would put patients in charge of their medical destiny, with doctors acting for them as servicing technicians. But there are two main ways of easing them.

The first is to set up standards for deciding who should be allowed to perform experimental surgery and for making sure that maximum scientific value is gained from each experiment. For example, with heart transplants – as the Board of Medicine of the American National Academy of Science has suggested[*] – the team should have done enough animal experiments to prove their surgical competence; they should be backed by skilled immunologists who demand the best possible tissue typing, etc.; patients should be observed with meticulous care – as in any scientific experiment – before and long after the operation; the results from every experiment should be shared among all teams as quickly as possible, even if this does mean setting up a complex communication network; and new experiments should not be done until the lessons of earlier ones have been absorbed.

Taking all these criteria, most heart transplants so far would not qualify and some would fall far short of them. With the Washkansky 'first', for example, though the blood groups of the donor and recipient were matched, there was only limited tissue matching – and that was not started until after the transfer began. On the other hand, what use are these criteria if a surgeon is convinced that a technique is not experimental? Who can force him

[*] *British Medical Journal*, 1 (1968), p. 762.

to obey the rules? As Dr Christian Barnard said of the Washkansky case, 'I was not doing an experiment; I was treating a sick patient.'

The other way of softening these dilemmas is of course to give the surgeons time and practice. Not to let them dash ahead without any restraints, but not to grind them to a halt either. 'Big' surgery has always had to advance over a growing pile of corpses before it could settle down to find its natural place in routine practice. For kidney transplants these early heroic days are now more or less over. Though some transplant surgeons still think kidney transplants should not yet be standard therapy, others say that even when the less successful cadaver organs are used they should be. Here, for example, is the leading British cadaver-kidney transplanter summing up a recent review of his last fifty-four transplants, fifty-one of which used cadaver organs:*

... the satisfactory quality of life of the majority of our patients has convinced us that cadaveric renal transplantation is valuable therapy. Our patients need no convincing: most had suffered severe symptoms of uraemia often complicated by hypertension, heart failure, peripheral neuritis, and pericarditis. To be restored from such a moribund condition to a normal life is worth while even if the ultimate prognosis is uncertain.

How soon the same kind of thing will be written about heart, liver and lung transplants it is too early to say. But it never will be unless transplant surgeons press on, and kill some patients in the process. The more transplant surgeons are attacked for being over-heroic, the more defensive they will feel, the more they will be limited to operating only on very sick patients and so the fewer patients will be saved. No surgeon waits until an appendix has burst and the patient has peritonitis before he is prepared to cut it out, because long ago he was given a vote of confidence that appendix operations were not over-heroic. One of the most difficult problems in the whole of transplantation is how to know when to give the transplant surgeons a similar vote – and so help to speed the time when the vote is truly justified by surgical success.

* R. Y. Calne, *et al.*, *British Medical Journal*, 1 (1968), pp. 404–6.

## Living donors

Every kidney transplanter would like to be able to do without living donors; no doctor likes deliberately mutilating a healthy person even to save the life of a dying one. But as long as kidneys from the living gave much better results than kidneys from the dead there was almost overwhelming pressure to use them. Today, though, the pressures to use live donors are rather different. Tissue typing has made it possible in principle to match a recipient with a dead donor as successfully as with a living relative. So if there were enough dead donors there should be no strong reasons for using live ones. The trouble is that there are not enough, and so if surgeons refuse to use live kidneys they are certain to let some of their patients die.

Imagine a young woman dying of kidney failure. She is a good candidate for a transplant, so she goes on a kidney machine to join the queue for a cadaver organ. She will probably have to stay in this queue for from four to six months before she reaches the top: in Europe until recently half of all kidney patients had to wait four months for a transplant, and the delay is getting longer. Even when she does reach the top she may have to wait months before a suitably matched donor dies: in Europe less than half of all kidney patients can expect to find a matched donor within a few months. This is a long wait, during which she has an appreciable chance of dying. Meanwhile one or more members of her family offer kidneys. How can their offer be refused? Faced with a situation like this, even surgeons who have strict 'cadaver-only' policies have been forced to break their own rules.

For a kidney donor there are physical and psychological risks. The physical risks start with the actual transplant operation. Up to the time of writing no donor has died of this, but the risk has been put at one death for each 2,000 transplants. Statistically, the first death must happen sooner or later. Once over the operation, the donor faces the fact that the human body was designed with two kidneys, and he had just given one of them away. He may, one day, badly need it back again. The risk of any kind of accident or illness affecting the one remaining kidney has been put at 7 per

1,000 donors. This is not, perhaps, an enormous risk, but it is not negligible either.

The psychological risks for the donor start with the true 'volunteer' nature of his sacrifice. A family will often select a donor from among themselves, even before a live transplant has been suggested. Such heavy pressures may be put upon that donor that he is made to feel he has murdered his sick relative if he refuses. In other cases a donor may 'volunteer' out of a sense of duty: 'I dread doing it but I ought to love my brother enough . . .' In yet other cases donors have been left feeling that they killed their relative when they have given a kidney but the transplant failed. Even if donation is truly and deeply voluntary at the time, recipients can be left feeling that they have a life-long debt they can never be free of, while donors can feel that the recipient should behave as *they* wish: 'After all it's *my* kidney she's got . . .'

These problems are taken so seriously that some transplanters – notably Professor J. Hamburger in Paris – put all potential donors through psychological screening in an attempt to clarify their motives for volunteering. But even this safeguard can still leave very nasty dilemmas, for while the family is being psychologically screened it must also be medically screened to find the most tissue-compatible donor. What happens when the most medically suitable donor is not the most psychologically suitable, and vice versa? Some transplanters have said that they put the psychological considerations first at all times, even if this means making the patient wait longer in the queue. But the temptation to ignore the psychological results when they clash with the medical one must be enormous. As far as I know no one has made the calculation, but it is a fair guess that the truly voluntary, psychologically safe, well-matched living donor must be a very rare specimen indeed – far rarer than a suitable dead donor.

## Dead donors

Taking organs from the dead has been compared to body-snatching and cannibalism and even called the 'final degradation of our Christian way of life'. Most people, however, seem to accept the idea and would rather have their organs used than have them burned or left to rot in the ground. For example, a Gallup poll in January 1968 found that 70 per cent of U.S. adults would let their hearts or other major organs be taken after death,* while in the same month an appeal to sign a 'Take my organs' consent form sent to 3·6 million British motorists by the Automobile Association produced a similar response. If willingness to donate were all, there should be no problems with spare hearts, lungs, livers and kidneys from the dead. In fact, of course, the problems are awesome.

The first difficulties are due to the fact that not everyone dies of the right causes – and when they do they have an awkward way of doing so in the wrong place. Medically, the perfect organ donor is a young person who dies of brain damage a few days after coming into a hospital very close to a transplant centre and who is known to be going to die so that the transplant team can be alerted to get his relative's permission to use his organs, to tissue-type him and so select and prepare the best recipient. The commonest perfect candidates are people dying of a brain tumour or haemorrhage. Because theirs is a brain death and a fairly quick one their other vital organs are unlikely to be damaged. Stroke victims may also make good donors: there are many of them, but all too often arteriosclerosis or a very slow, lingering death will have damaged the other organs. Major heart surgery is another good source (but not, of course, for hearts or lungs) since everyone can prepare for the worst beforehand.

This leaves most people as imperfect candidates. Except for primary brain tumours, which do not spread malignant cells

---

* Of the 23 per cent who said 'no' (the others were 'don't knows') the commonest reasons were age and health ('My organs are about worn out and my heart's not what it used to be. They'd be taking a chance with me'); moral or religious objections to mutilation ('I've always been taught that the human body should not be touched after death – it's just not right'); and doubts about the success of transplants ('You can't tamper with human life like that. These transplants will perhaps stall a death a week or a month but I don't believe they'll ever get a man back on his feet again'). *New York Times*, 17 January 1968.

through the body, cancer deaths are ruled out: in the early days many transplant recipients contracted cancer from their new organs. Infection, high blood pressure and any prolonged death rule out most other donors. So should old age, because of the general wear and tear on organs, but owing to the acute shortage of donors they sometimes have to be used as a rather desperate second-best.

One of the biggest potential sources of spare parts are victims of accidents or violence, because many of them die from head injuries which leave their other organs intact. If they take a few hours or days to die in hospital they are perfect candidates; but all too often they die quickly – leaving those sound organs on a motorway in an unidentified body or being transported in an ambulance that is not heading as fast as possible for a transplant centre. Building transplant centres in accident hospitals is one of the most rational, if gruesome, answers to this. But it still leaves the problem that the young, healthy-looking accident victim who dies quickly may have diseased organs which only his own doctor – perhaps asleep at home, a hundred miles away – knows about. Many kidney-transplant patients have died because their new organs turned out to be infected.

Once the medical criteria have been satisfied, time is of the essence. To minimize ischaemic damage the organs have to be cut out of the dead donor as soon after death as possible, which brings us to the twin problems that arouse the deepest misgivings. How is it known that the donor is dead? And with the transplant team waiting, will the doctors give up trying to save the donor earlier than they would otherwise?

Many surgeons assert that these questions are a sensational red herring. They claim that they have existed ever since there were life-supporting machines to switch off; that doctors are accustomed to judging the moment of death; and are certainly used to making these judgements on their own. Their case is strengthened by the fact that in hospitals machines are scarce. Often they must be switched off in favour of another patient judged to have a better chance. But other surgeons insist that when a transplant patient is waiting there are strong temptations to switch machines off early; perhaps even before the potential donor is traditionally dead. And of course the fear in everyone's mind is that as long as doctors and their patients do not all recognize the same definition of death, the delicate balance of trust between them may be upset. We all want to feel that we shall be given maximum chances of recovery.

In the old days it was easy to define death: you were dead when your heart stopped. Nowadays it is far more difficult, because every day in hospitals hearts are restarted by cardiac massage and electric shock. If the heart is restarted within the four to five minutes that it usually takes for the brain to be so deprived of oxygen that it is irreversibly damaged, and then continues on its own, clearly the patient is not dead and no doctor would say he was. But how often does one restart a heart that keeps on stopping, and what is irreversible damage? To make things more difficult, the ideal donor, with his severe brain wound, is more than likely to be on an artificial respirator. With it and diligent nursing he can be kept breathing and his heart kept pumping often indefinitely. If there is a reasonable chance of recovery, again he clearly ceases to be a donor. But what is 'reasonable'? And what does 'recovery' mean – life with a normal brain or a semi-idiot existence?

If the doctors are not sure they will almost certainly give the patient the benefit of the doubt and keep him on the respirator. At some time, though, they may decide to abandon him as a lost cause by giving up their extraordinary measures to prolong his vegetable existence, and switching the machine off. In some cases this decision can be made easily, very early on: if the patient has half his skull smashed in so that his brain is irreparably and irreversibly damaged he is essentially dead – even though his heart is still going. Though heart-stop is the traditional sign of death it is the brain that determines whether life continues. In most cases there is neither the certainty of recovery, nor of irreversible brain damage, but the uneasy no-man's-land of guesswork. Meanwhile the doctors looking after the 'dying' man know that he is a potential donor, and that by lifting a telephone they could alert a transplant team and give a hope of life to a waiting transplant patient. Can they resist the pressures to make that call before the 'proper' time? It has often been suggested that they should and must; that they should act only when they are sure that their patient is dead or there is no further hope of saving his life. Unfortunately, however, because the transplant must be done very soon after death and it takes time for the transplant team to get *their* patient ready, the call has to go out before death. There also has to be time to get the permission of a coroner and the donor's family to remove the organs. So everyone must know in advance of death: the pressures to switch off early are magnified.

It seems that our eventual definition of death will ignore the heart in favour of the brain and nervous system; that we may reach a point where a man whose heart is still beating may be declared dead. In some centres organs have already been removed from donors under these circumstances. Other surgeons still shudder at the idea.

The criteria for brain death are not as simple as the stopping of a heart. Indeed, no one has finally agreed on them. The nearest we have got to a definitive statement was produced at an international meeting of cardiologists, surgeons, immunologists, neurologists, etc. organized by C.I.O.M.S. (Conseil International des Organisations des Médicales Sciences) in June 1968. The relevant part of the statement runs:

Cerebral function must have completely and irreversibly ceased. The criteria for cessation of cerebral function are as follows:

(a) Loss of all response to the environment;
(b) Complete loss of reflexes and muscle tone;
(c) Absence of spontaneous respiration;
(d) Massive drop in arterial blood pressure when not artificially maintained;
(e) An absolutely linear electro-encephalographic tracing recorded under the best technical conditions, even with stimulation of the brain.

These criteria are not valid for young children, or for subjects in hypothermic states (that is, extreme chilling) or with acute toxic conditions (which they might have after attempted suicide with drugs).*

Obviously the ghoulishness of this problem can be relieved only if there is less urgency about removing organs. Some transplanters see themselves with horror as vultures hovering around the dying donor – and the public reputation of transplantation suffers from this image. We need time to preserve the decencies, but we also need organs in the best possible condition.

Better life-supporting machines would help. Then, however long a donor took to die, his organs would not suffer. The logical extension of this is that organs could be preserved in living vegetables, long since pronounced dead. But most people would violently object to this: they would feel, with Professor Roy Calne, that it is wrong to 'meddle with the donor for the purposes of improving the transplant'.

A quite different ethic has been suggested by the heart surgeon W. J. Dempster, of the Royal Postgraduate Medical School, London.† He argues that instead of worrying about patients in a hopeless condition being switched off too soon we should be far more worried about them being kept 'alive' too long. We should

*To see why consider the following extreme case abstracted from the *British Medical Journal*, 1 (1963), pp. 1315–17.

In March 1962 a five-year-old boy fell into a partly frozen Norwegian river and lay there with his head beneath the water for at least 22 minutes. When a doctor reached him he was 'apparently dead, the pupils were widely dilated, and the skin blue-white'. After two and a half hours of mouth-to-mouth respiration, heart massage, drug injections and blood transfusion the boy's heart and breathing started to work again spontaneously. Five times during the next twenty-four hours his breathing stopped again but was restored by an artificial respirator. His blood was totally exchanged by a transfusion and he was fed intravenously for a week. On the tenth day after his accident he had recovered sufficiently to recognize his mother and answer 'yes' and 'no'. Then he relapsed into total unconsciousness for five weeks and for part of this time had no measurable brain function at all: in the jargon he was 'decerebrate'. However, during the seventh week his vision returned and six months later he was back at home, a normal child again except only that his finer finger movements were a little clumsy and his area of vision was slightly reduced.

†'Legalize euthanasia for transplants', *Medical Tribune* (U.K. edition), 9 May 1968.

worry for the prolonged agony of their families. And we should worry for the eventual recipient of their organs, damaged in this long, slow but inevitable death:

The quixotic humbug of transplanting severely damaged organs is not in the best interests of the recipients. The rights of the dying donor have been the cause of anxiety in the past, but, accepting that the donor is a hopeless case, we must surely now consider the rights of the prospective recipients ... It is, therefore, absurd to introduce 'safeguards' which would exclude the transplant surgeon from the decision to switch off the respirator. Indeed, in formulating any such safeguards are we really warning the public that an enthusiatic transplanter would prematurely end the life of any given hopeless respirator case? Whichever way we care to look at these safeguards, they seem to reflect the nightmarish suspicions of a sick society. The transplanter surgeon must play his part in the collective declaration to end the artificial life of a hopeless respirator case.

However, Dempster adds that neutral arbitrating doctors, including neurologists, must agree that 'the point of no return has been reached and that the patient would never recover'.

Once the donor is declared dead it is possible to preserve his organs within the body. If this is done, the transplant team and recipient need not be waiting in the operating theatre for the moment of his death. It is already fairly common practice to massage the dead donor's heart and apply artificial respiration, thus keeping his organs perfused with oxygen-enriched blood. Some surgeons go further, and chill the donor's body by connecting an artery and vein to a cooling pump. And in the near future it should be possible to remove the organs and store them for days, weeks, or possibly even months, thus finally severing the all-too-intimate link between donor and recipient.

Indeed, there is a case for saying that cadaver transplants, especially of the heart, should have been delayed until organ banking was feasible. The order in which medical developments take place is often just as important – for technical *and* ethical reasons – as whether they take place at all. If organ banking had come before transplanting, much grief and anger might have been spared. But would anyone have developed organ banks before organ transplanting made them necessary? Probably not, but the point is that perhaps someone should have told them to. The need for an overall medical research policy which includes such forward-looking considerations is clear.

Meanwhile, transplanters are desperately anxious not to miss any suitable organs which become available. At present they miss thousands each year because of the legal requirements which must be fulfilled before an organ can be removed for grafting.

Under English law (Human Tissue Act 1961) a dead man has no rights in his body. Even if he has bequeathed it or parts of it for medical use, after death his next of kin are not bound to honour his wishes. So if a man dies suddenly the transplanter must find his next of kin, break the news to them and, while they are in a state of grief and shock, get them to sign a consent form. This can take precious time and is brutal for everybody concerned.

What transplanters want is to change the law so that they can take organs from any dead body unless there is good reason to believe that the donor himself would not have wished it. In other words, they want a blanket permission which throws the onus of refusing organs on to their original owner. This 'silence implies consent' ruling already exists in many countries.* In Britain, though, there is a good chance an exactly opposite rule may be established, whereby organs can only be taken from a register of voluntary donors.†

* In the U.S.A. about half of all states have introduced laws which, essentially, make the provisions the transplanters want, though in detail rulings vary widely. But there are strong moves afoot to persuade all states to adopt the same laws. In France the legal position is curiously reversed. The law allows an organ to be removed from the dead without any need to get family permission, but it does not allow kidneys from living volunteers. Surgical operations on humans are only permitted if they will be medically useful to the patient himself. Transplanters are attempting to change these laws too.

† In Britain an advisory group set up to study the question reported in July 1969 that while six members favoured giving the transplanters what they want, five felt that organs should be used only when the donor had definitely indicated his willingness while alive, and that a single public, central register of 'consenters' should be established. Transplanters would

But both these proposals ignore one very important factor. Although most people tell pollsters that they would like their organs used, very few can have thought deeply about what this might mean to their families. As a society we have already reduced the ritual of mourning to a psychologically dangerous minimum. How are we really going to feel when other people walk about with our dead children's or husband's or wife's living hearts? At the moment families are at least asked and can refuse if they can bear the guilt implicit in doing so – in refusing the chance of life to somebody else. Most give permission, as transplanters are only too keen to emphasize. But as far as I know there have been no compassionate, much less psychiatric, follow-ups of the resolution of grief in these cases. I, for one, will never forgive the doctors involved in the first British heart transplant, who took the donor's heart before his pregnant wife knew of his death. Or the society which almost completely ignored this in the excitement of their immediate success.

## The spare-part market

Already there is an acute shortage of donor organs, and patients die because of it. But when transplanting kidneys, hearts, livers and lungs (in roughly that order) becomes routinely successful what is going to happen then? Will there be even more acute shortages with all the unpleasant problems of selecting and rejecting patients? Or will medicine and society be able to accept calmly that this new technology is limited, is not a universal gift of immortality, while trying to expand the limits by setting up an efficient service to get all available donors to the customers who need their parts? These are two of the most formidable challenges that transplanting will present us with as it becomes a success.

It is very hard to estimate whether there will be enough organs

---

have to check that any potential donor was on this register before taking his organs. Because of this sharp division it is likely that the law will stay as it is until the public clearly expresses its wishes, while a trial computerized 'consenters register' will be set up.

to go round. Looking more than ten years ahead is rather like trying to predict today's jet airliner travel – with all its success *and*
its failures to provide cheap flights for all – in the days of Blériot.
Disease patterns will change dramatically, slashing the demand
for spare parts – and the supply of donors. Criteria for selecting
patients will change: while the new disease patterns lead to later
deaths, surgeons might switch to transplanting younger and otherwise healthy people. Nevertheless, we can only work from what
we know now, so it is worth looking at today's supply-and-
demand estimates. We shall take the demand for organs first.*

### Kidneys

As we saw in Chapter 9, the annual demand in the U.K. is
between 2,000 and 3,000 (assuming that transplantation entirely
supersedes dialysis). In 1967 there were about 100 kidney
transplants in Britain, but the National Health Service plans to
increase this figure to 600 in the next few years. The remaining
candidates for kidney transplants will die or go on kidney
machines.

### Hearts

Similarly, from Chapter 9, the annual demand in the U.S.A.
for total heart replacement is about 86,000 for people facing
immediate death, plus 46,000 if the cardiac-cripple population
gets new hearts over a ten-year period. Worked out on a similar
basis there would be a U.K. demand of about 22,000 plus 11,000
a year. Donald Longmore has suggested that a *further* 30,000
patients in the U.K. could benefit from heart-plus-lung grafts.

---

* For kidneys and livers detailed estimates have been made, but only for
the U.S.: Nathan Couch, 'Supply and Demand in Kidney and Liver
Transplantation: a Statistical Survey', *Transplantation*, 4 (1966), pp. 587–95.
I have used these estimates here with two modifications. First, Couch's
figures are for 1963 and I have adjusted them for the larger 1968 U.S.
population. Second, when applying them to the U.K. I have adjusted for
the smaller population by dividing them by four.

## Livers

Deaths from liver failure are roughly the same as for the kidney. The two major causes are cancer of the liver, gall bladder and bile ducts (about 1,600 in the U.K.) and cirrhosis (5,400). But for three reasons there are many fewer good transplant candidates. Cancer and cirrhosis are predominantly diseases of the old; most cancer victims would not be suitable because the malignant cells would have spread through the body; and about three-quarters of the cirrhosis victims are chronic alcoholics. Even on easy medical regimes alcoholics fail to follow advice and die sooner than normal patients: as long as transplant patients have to be selected very few surgeons would take them on. Taking all this into account, and limiting the age range to under sixty there are about 600 to 1,000 good candidates a year.

## Lungs

In Britain there are nearly 30,000 lung cancer deaths a year and 31,000 deaths from bronchitis. There are also tens of thousands living a very restricted life as bronchitic respiratory cripples. How many of these could be saved by lung grafts no one can really predict. Many of the cancer victims might have to be excluded, and many of the bronchitics will be too old and generally worn out. As with heart disease, one can also expect major reductions in lung disease when (or if) preventive social medicine gets its teeth into the problem. If these reductions merely delay the onset of the diseases, as seems likely, there would still be a big drop in demand because candidates would tend to have other serious disorders. At present a very rough estimate of the demand for lung grafts might be half the death-rate, or, say, 30,000 a year.

Now what about the supply side of the market? Are there enough suitable donors to match this demand?

The ideal donors are people who die of a subarachnoid haemorrhage – a burst blood vessel in the membranes linking the brain. They die fairly quickly, usually in hospital, and their other organs are very rarely damaged. In the U.K. there are about 1,600 such deaths a year. If all of them did die in hospitals near transplant centres they could provide 1,600 livers plus about 2,800 kidneys (for surgical reasons not everyone can donate two kidneys). These donors alone could more than satisfy the demand for livers, just about meet the demand for kidneys and could provide roughly 1,500 hearts and lung-pairs.

Then there are the other 'natural' brain deaths: primary brain tumours, strokes, cerebral haemorrhage and other central nervous system lesions. Together these add up to about 10,000 deaths a year, of which about 5,000 might be good organ sources. Lastly there are victims of road accidents (about 7,500 a year) and other accidental deaths and violence (10,000 a year). No one seems to know how many of these victims die from pure head injuries without damage to their other organs, or how many die just before or soon after arriving in hospital. But taking these last two groups together something like 10,000 potential donors for heart and lungs (as well as kidneys and livers) seems reasonable. Longmore has estimated there are between 7,000 and 14,000 potential heart or heart-plus-lung donors a year.

To summarize: if there were a perfect donor supply system there would probably be no shortage of spare livers and kidneys, but when heart and lung transplanting becomes routine there will almost certainly be an absolute limit set by the scarcity of donors.

This does not appear to raise any unique problems. Of course, it means that transplant patients must be selected and others left to die. But there is nothing new in this, as we saw in the last chapter. Some commentators have suggested that there could be one special problem: a black market in organs. From the analogy that in some countries shortages of blood have led to payments to blood donors, they predict that a few rich people who cannot get a spare organ will pay a high price for live donor kidneys or dead men's hearts while a few poor ones will be prepared to sell. I believe this could not possibly happen. Every transplanter is absolutely adamant that it should not, and it is almost inconceivable that their attitudes will change in the foreseeable future. In that case, the black market would have to be an underground movement, like illegal abortions. But how do you set up a criminal transplant centre, with operating theatres, skilled surgeons and all the paraphernalia of getting, typing and cooling donor parts? It is a trifle far-fetched, though it might be a good idea – if the majority thought that a black market in organs was a grossly unethical idea – to take any necessary legal steps now to prevent it.

Meanwhile, there is the further obvious limitation that there are far too few transplant surgeons, or funds to build and equip centres for them. In the short term this means that transplant candidates die and, equally serious, that transplant operations often take up scarce operating theatre time which could be used for other needy patients. For example, in the week following the first British heart transplant at the National Heart Hospital there were no major heart operations, though in a normal week there are ten or so. The reason was that the recipient, Frederick West, was sealed off in the twin-theatre operating suite because this was the only place in the hospital that could be made sufficiently sterile.

How this mainly socio-economic limit will be resolved is anybody's guess. Society might vote massive funds for transplantation, as it has for other growth technologies. Or it might choose instead to spend its money elsewhere. This, however, is not the place to discuss such choices. They are not peculiar to transplanting but are so general and fundamental that they deserve a chapter to themselves – the final one. All one might say now is that there is no inherent reason why transplantation should not expand right up to the limits set by the supply of donors; but that, if we choose that it should, there are further, immense challenges to face.

## A world tissue service

A man is dying slowly of incurable heart disease in a London heart hospital. He is a good transplant candidate, and so a tiny tissue sample is removed and sent to one of the handful of tissue-typing centres in the world. Luckily there is one near by, for the sample must reach it within four hours. Meanwhile, his doctors put out calls to friendly neurological and accident units saying they are on the look-out for a donor. One of them replies that a car crash victim came in during the night mortally wounded in the head. He is on a respirator, his next of kin have agreed that he could be a donor (after a six-hour search for them and a harrowing discussion), and his tissues have been sent for typing. Then the results come in and the victim dies. His tissues do not match those of the heart patient, so he is sent down to the mortuary intact. The heart patient dies the next day. Meanwhile in a Manchester hospital a girl with kidney disease dies, and in a Paris hospital a man with cirrhosis of the liver dies, because there were no donors available. Both of them were perfect tissue matches with the accident victim. And meanwhile in a Stockholm neurological unit a doctor decides a young woman with a bullet in her brain is past any hope of recovery and reluctantly switches off her respirator. Her tissues were a perfect match for the London heart patient. Two sets of organs not used and (at least) three lives not prolonged – all because of imperfect communications. As Donald Longmore has written * 'This is not even the philosophy of the horse-and-buggy, it belongs to the barefoot-runner-with-cleft-stick era. If transplantation surgery is to turn promise into performance, to apply even our present knowledge and skill to the saving of life, something much better will have to be organized.'

That something is beginning to be done. In Europe, hospitals in Holland, Germany and Belgium are linked to the Eurotransplant centre at Leiden; another integrated network is being nurtured to cover Denmark, Sweden and Norway; and in Britain a central tissue-typing service has been set up in Bristol. In the U.S.A. there are networks centred on Los Angeles and Minneapolis and plans for similar schemes on the east coast. Ultimately, these co-operatives may expand into super-networks covering the Americas, Europe and Africa, and Asia and Oceania, linked to give a world service.

* 'Implants or Transplants?', *Science Journal* (February 1968). See also his book *Spare Parts for Man* (Aldus Books, London, 1968).

If it succeeds, the world tissue service will be an extraordinary organization. It will have to process a vast amount of information and convey it to and from thousands of medical centres at great speed, using every resource of modern communications, from telephones to large computer stores and satellite links. It will have to store, in a hierarchy of regional and super-regional centres, data on every patient waiting for a transplant. This information will include not only the patient's tissue group and organ needs but more problematical facts about medical (and perhaps social) *priorities* for an organ (nearness to death, length of time spent waiting on a kidney machine, age, strength of motivation to make the transplant succeed, etc.). All these data will have to be updated daily. Then the tissue service will have to do the same for potential donors, as they come into hospitals to die or undergo high-risk operations. To make this side of the business work best, donors will have to be recruited while they are still healthy. Longmore has suggested that to help recruitment there should be a rule that anyone who volunteers his organs for use after his death should automatically get a higher priority rating if he ever becomes a transplant candidate. To help speedy removal of his organs after death without seeking his relatives' permission, donors would have to be recognizable – perhaps by a 'D' tattooed under an arm. 'To some, no doubt, it will be mainly a status symbol. To many, though, it will be something more profound: the mark of the person who cares.'

As the streams of data on patients and donors come in, the computers will be brought into action by two kinds of emergency: the sudden need for, or the sudden availability of, an organ. They will then calculate the best donor-recipient matches, taking not only tissue compatability into account but urgency of need and, not least, the travel time between the donor parts and recipients. This done they will automatically warn the recipients' hospitals, rout the parts and alert the transport system – taxi, plane, ambulance or whatever. With a large network the whole system would have to be very flexible and fast: as a heart from Berlin is flown to London, London may be told that a fresher, better-matched heart is being flown in from Paris, while another London hospital is warned to expect the Berlin heart.

Until complex organs can be stored for more than an hour or two this kind of system is obviously limited, so limited indeed that transplant centres might even have to be built near airports. But organ banking looks as if it will come. There is no point in giving details of current achievements, as they will be outdated too quickly. But as a pointer to the future, in March 1968 Dr F. O. Belzer of the University of California described a kidney preserver that he hopes will keep kidneys in good condition for three days and has already stored a kidney for seventeen hours before it was grafted successfully into a human patient. Animal grafts with kidneys stored for three days have already succeeded. The machine works by pumping chilled, oxygenated and fortified plasma through the kidney. As for hearts, a French group has recently claimed to have restarted hearts taken from cadavers forty-five minutes after death and hearts which have been stored for forty-eight hours after removal. The technique here, apart from chilling and fluid perfusion, is to pump the fluid through the heart in pulses at the heart's natural resting rate – sensed by inserting an electrode into the heart's natural pacemaker.

Gruesome as they may seem to some, banks like these must make transplanting less disturbing. For apart from shrinking the distance between donors and recipients – so saving many lives – they will also expand the time between a donor's death and the grafting of his organs into a recipient. Organ banks will also make donors and recipients truly anonymous, as they are with blood banks today – if, that is, those foot-in-the-door pressmen let them be so.

A world tissue service and banking system will not be easy or cheap to establish, but it is not beyond our wits or resources to set one up if we want to. Prototypes already exist in the vast and rapid global data networks of the World Weather Watch, or blood bank services. The question is whether we want to, or rather, if we do not say what we want, whether the service will grow more slowly than we would like or, on the other hand, will the transplant lobby push it faster than we would like? It should be up to us.

Whether or not we develop organ banks and world tissue services, it appears that there might always be an overall spare-part shortage. As Sir Peter Medawar has said, there is a basic paradox here: as our medical armoury grows more powerful we shall prolong the lives of donors as well as recipients, in many cases by making *them* recipients too. In the long run we may have to break out of this vicious circle by going back to where human transplanting began – with animal donors.

## Animal organs

In modern times the first animal-to-man transplant (heterograft) was done by Dr Claude Hitchcock at the Minneapolis Medical Research Foundation on 16 February 1963. He grafted a baboon kidney into the thigh of a sixty-five-year-old woman dying of irreversible kidney failure. Three days earlier she had received a human kidney, but it had failed. There were no other human donors for a second attempt, so her only hope – and it was known to be almost nil – was the baboon kidney. Though it worked for a short time after the transplant, by the fourth day it was failing, it was removed, and the patient died eight days later.

The leading pioneer of animal organ grafting is Professor Keith Reemtsma. Between October 1963 and February 1964, in New Orleans, he grafted two Rhesus-monkey kidneys into a woman and chimpanzee kidneys into six patients. Again, all the patients were on the verge of death and no human donors or kidney machines were available. All the grafts failed, but one patient – the first to get a chimpanzee kidney – lived for sixty-three days until he died of pneumonia, largely because of heavy immunosuppression. On autopsy there was no sign of rejection. Since then there have been a few abortive animal-to-human heart grafts and sporadic but unsuccessful attempts to graft animal livers into humans.

Whether animal organs can relieve all the scarcity and moral difficulties of human donors by becoming our major spare-part source is entirely open. To a large extent it is up to what immunologists discover, and social acceptance and effort.

Biologically speaking, it does seem quite likely. For one thing, though as a general rule animals must be immunologically more different from men than men are from each other, heterografters like Reemtsma have argued that quite possibly some primates will match some humans better than many humans match each other. With more research on primate immunology and genetics it might well be possible to breed special 'man-matching' strains. Should we back such research massively? Then there are the hopeful prospects of A.L.S. As Sir Peter Medawar has said, biologists are now getting tissues to take *across* species better than they could get them to take between animals of the *same* species fifteen years ago.

When (or if) the animal/human rejection barriers are broken down there is the problem of breeding the right animals. Chimpanzees and baboons have kidneys quite similar to those of humans, but their hearts might be too small. Probably the bigger primates, such as the orang-utan, would have to be used; but they are scarce and hard to breed in captivity. Anatomically, pigs might even be the best donors of hearts and the other organs, and it should not be too difficult to breed pigs more or less the same size as humans. No one really knows. There is an enormous amount of applied biology to do here, and human experimenting too, before we know just which animal parts would adapt best to humans. And when that is done we shall have to breed special herds of them: animal factories to feed us spare parts on the conveyor belt. Who knows how we shall feel when we drive past one of these grunting organ banks, or what living with an animal organ will do to our images of ourselves? Ought there not to be a widespread psychological research programme to answer such questions *before* the biocrats show that it is technically feasible and we get committed to the idea? To put it mildly, this way out of the transplantation paradox will not be as easy as some optimists suggest. And we should not let it be too easy: unless we take great care we may in a deep sense find ourselves the losers.

Finally, there is one set of questions by which we must judge the transplanters and all those other spare-part surgeons with their mechanical organs. Do we want them at all? Do we really want them to succeed as gloriously as they would wish? As the stars of modern expensive curative surgery, are they leading us along an attractive but wrong road? It is not just a question of affording them – a question of means – though every pound spent on them is a pound not spent elsewhere. It is also a question of ends. When they reach their bright future will they have become the props of a dying race?

These questions are so overwhelmingly important that I have left them to the last chapter to discuss. In the meantime, let the gist of what I am trying to say come out of the words of one of Britain's most eminent physicians, Sir George Pickering, talking about transplants.

The present goal of medicine seems to be indefinite life, perhaps in the end with somebody else's heart or liver, somebody else's arteries but not with somebody else's brain. Should transplants succeed, those with senile brains will form an ever-increasing fraction of the inhabitants of the earth. I find this a terrifying prospect.

# 11. PRIORITIES

The greatest of all the challenges with which medical progress is confronting us stems from a simple fact. We cannot afford it. As our medical 'cure power' rapidly outstrips our 'pay power' medicine is becoming like a branch of Fortnum and Mason set in a slum ghetto. Each year the shelves of the medical shop are stacked with more and more goodies – with surer ways of pushing back death or banishing sickness – yet each year, too, the customers are finding themselves less and less able to afford all the tantalizing promises on offer.

As a result there is a swelling insistence that what goes on sale must be rationalized. If we cannot afford it all, we have to decide what expensive luxuries should be restricted so that everyone can have the necessities. But this rationalization, which most people believe must be made if medicine is to avoid sinking into a catastrophic crisis of practical and moral bankruptcy, will be fearfully difficult to conduct. For it means not only treading on the toes of almost every vested interest in the medical profession; not only choosing which lives we cannot afford to save and which we can; it means digging down to the core of medical practice and asking, What is medicine for?

The founders of Britain's National Health Service thought they knew what medicine was for. They insisted that the whole population had a basic right of access to the best modern medical care and that this could be achieved only by making it free. Given that, they hoped that with good medical care everyone would become healthier. Like Lenin's state, medicine would gradually wither away. Accordingly, they predicted that while the N.H.S. would cost £170 million in its first year (1945), despite a growing population it would still cost about the same twenty years later. Instead the 1965 N.H.S. bill was £1,308 million, and by 1970 had quadrupled since 1950. While much of this soaring increase is due to general inflation – the proportion of National Income taken by the N.H.S. has risen only from 4·4 to 5·4 per cent in the last fifteen years – it clearly has not been enough. Though death-rates have declined and the population is generally healthier, the gap between promise and achievement has grown enormous. And everyone knows it. Whether or not medicine is doing well, the outcry that it has to do better is swelling into a deafening chorus.

What has gone wrong? Why, in every advanced country, is medicine failing to meet our hopes for it, despite vast inputs of manpower and money?

One of the key reasons is that the pattern of disease has changed significantly and rapidly. While our grandparents were physically energetic but plagued by infections, we are now scourged by the social, degenerative diseases of middle and old age. Heart disease, cancer and stroke are now the three leading killers in virtually every northern-hemisphere country, followed by pneumonia and accidents plus violence. Though this swing has been going on steadily through the century, Figure 11.1 shows just how much it has accelerated in young and middle-aged people in the last twenty years.

The result is that medicine has largely lost the huge bonus given to it by the relatively cheap drug control of disease – by the vaccines of mass-immunization programmes and by the 'miracle' drugs such as antibiotics. Most of that huge drop in deaths between the mid-1940s and mid-1950s was due to these cheap weapons. Instead it is now facing three new pressures on its scarce resources.

First, to try to prevent the deaths from the new scourges usually demands highly specialized hospital treatment with expensive man-power and equipment. For every heart transplant that highlights this fact, there are thousands who have less spectacular but hardly less expensive kinds of open-heart surgery, or who live for weeks in intensive-care heart units monopolizing thousands of pounds-worth of sophisticated gadgets and often more than two medical staff per patient. Second, the great surge in chronic diseases has brought a tremendous load on medicine. These diseases tend to need complex, long-term treatment instead of the specific, short-term therapy which infections need; and because they are usually too complex for a single physician to cope with they need the integrated team-work of many different specialities. All this adds to the complexity – and the cost – of treatment and care. Third, with the explosive development of powerful new drugs and death-averting medicine and surgery, death-rates have been cut and the population has aged. Half the population now live into their seventies, often with a mixture of chronic degenerative diseases and mental ill health, and though by and large we do it horrifyingly badly they have to be cared for. Unfortunately for medical resources, old people on average need medical treatment twice as frequently, with each treatment lasting twice as long, as do the rest of the population.

Because of these trends the total cost of hospital treatment – itself a major part of all health expenditure – is rising rapidly. In the U.S.A. it soared by 16·5 per cent in 1966 and a recent high-level study* has estimated that in the decade ending 1975 it will have risen by 250 per cent. In Britain the rise is slower but is still more rapid than total health spending or National Income: 11·5 per cent in 1965–6 and 7·5 per cent in 1966–7 (this smaller rise being largely due to the wage freeze). The hospitals, in short, are taking a continually larger slice of the health budget; and when one considers the antiquated, depressing, understaffed, queue-filled state of the vast majority of hospitals it is obvious that the slice is still not large enough.

And yet it is a curious fact that despite these vast investments in the specialized, high-technology medicine that the hospitals broadly represent – the kind of medicine that deals with the dramatic illnesses and death-postponement, where action is urgent and the latest skills are concentrated – death-rates have now more or less stopped declining except among the very young and the old (see Figure 11.2).

* *Report of the National Advisory Committee on Health Manpower*, vol. 1 (U.S. Government Printing Office, 1967).

FIGURE 11.1

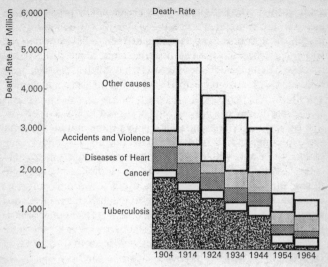

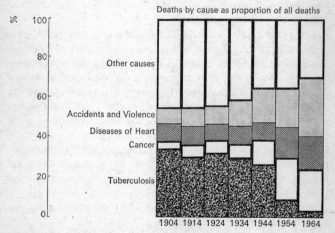

Leading causes of death in the 15–44 age range, England & Wales. The top graph gives death-rates per million living; the lower the deaths by cause as a proportion of all deaths.

Similarly, in the U.S.A. there has been a barely perceptible increase in life expectancy since 1954. This is not the fault of the hospitals, of course. The fact is that the 'social' diseases are tightening their grip as fast as modern scientific medicine is trying to loosen it. But what it does suggest is that much expensive, high-technology, curative medicine is becoming an increasingly poor investment.

The second key reason for rising costs – or rather for pressure that costs should rise even faster – is that more and more people are expecting, indeed demanding, treatment for 'minor' complaints. In Britain, with its free health service, it is very rare not to get at least adequate care for an acute or life-threatening condition. But it is increasingly common not to get any care – let alone the best – for life-diminishing afflictions. This is one of the great paradoxes of medical progress: as a population becomes healthier, as its living and medical standards improve, the standards of what it means to be 'healthy' rise so that the total of reported illness actually increases. Victorian parents philosophically accepted the death of several of their children; we now jibe that scientists have not found a cure for the common cold.

This revolution of expectations is not just a scourge of mass hypochondria. Many people have pointed out that ill health is like an iceberg. Sticking out of the water are the conspicuous diseases – from, say, fulminating lung cancer at the peak to a severe bout of influenza down at the water line – which people are well aware of and do get treatment for. But beneath the waves there is a large submerged part of the iceberg consisting of all the ills which for various reasons people do not get treated. The point is that while rising expectations are lifting this hidden part higher out of the water, medical progress is showing that for good reasons it ought to be lifted out, because by no means all these hidden diseases are tolerable or trivial.

Several attempts have been made to measure how big the hidden part of the iceberg is. One of the best recent ones carried out in Britain was The Bermondsey Health Survey of 2,500 adults and their 700 children conducted by

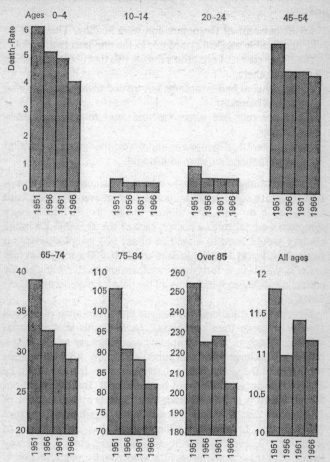

FIGURE 11.2 Death-rates in Britain

(Death-rates in each age group per 1,000 population)

Professor W. Butterfield. What it found can be summarized as follows:

(a) 12 per cent of the population were healthy. They had no noticeable medical symptoms in the previous two weeks.

(b) 30 per cent had symptoms during this time, but did nothing about them.

(c) 35 per cent had symptoms but treated themselves by going to a pharmacist.

(d) 15 per cent had symptoms and went to a general practitioner.

(e) 8 per cent had symptoms and were either going to hospital as outpatients or were in hospital.

Extrapolating these figures to take in the whole country reveals that there are some 1,000 million symptom-events a year, with action taken on only 600 million. It also reveals that if every symptom were taken to a doctor, each of the 23,000 G.P.s in the country would have to attend to about 900 symptom-events a week. As it is, on average doctors attend to 200 symptom-events a week, which in a forty-hour week means twelve minutes for each clinical decision, including all the time for paper-work, home visits, etc.

Of course, it is ludicrous to suggest that the number of doctors should be more than quadrupled to meet this vast potential demand. Many of these symptoms are genuinely minor, like a cold or headache or menstrual pains. With many others (neuralgia for instance) the victim has probably long since learned that if he did go to a doctor little could be done for it. But many are not trivial. There is strong evidence that there *is* a vast amount of fairly major disability and ill health in the population that is not being treated. In some cases the 'patient' has just got used to living with it, even though he may be living quite a long way below par: the chronic bronchitic who coughs and spits and is often short of breath is a typical example. In others he just has not realized his medical 'rights': before the National Health Service, uncorrected bad vision and partial deafness or the loss of all one's teeth were commonly accepted as a normal part of ageing. Now we think not, but many old people still do not realize it: as Professor Peter Townsend has pointed out, 95 per cent of the medical facilities for the aged are used by only 5 per cent of the age group. When the others realize what is available the iceberg will be forced quite a way out of the water. In many cases the 'patient' may genuinely not recognize that he has a medical abnormality. Though often this does not matter, because the abnormality is just a trivial deviation from the normal, very often what in fact he has are the early 'silent' stages of the degenerative diseases – diseases which develop insidiously and all too often go unrecognized until it is too late to treat them effectively. We shall see in a moment what screening programmes to pick up these early signs might do to change this. But for now, in case anyone thinks 'vast amount' is an exaggeration for these untreated, hidden diseases or that they are not one of the major priorities for medicine to tackle, consider the two British estimates of their extent shown in Figures 11.3 and 11.4.

If all these shortcomings are true of Britain they are doubly true – and the health crisis is doubly critical – in countries where health is not free and open to all. Consider the U.S.A. In America, health spending has nearly quintupled over the last twenty years, from $11,000 million in 1949 to about $50,000 million in 1967–8, and has risen twice as fast as the cost of living. In other words there has been a real rise, with health now taking 6 per cent of the Gross National Product (the highest proportion in the world). With 3,000,000 workers, the health industry is the third largest in the nation.

Until 1966–7, when Medicare and Medicaid were introduced, three quarters of all American medicine was private; now two thirds of it is. So medical care largely depends on health insurance schemes. But though the proportion of people with some form of insurance protection has soared from roughly 10 per cent in 1940 to 80 per cent in 1965, on average it only covers a third of the family's health care bill. With the high cost of care, most low-income families delay going to a doctor until their condition becomes intolerable – by which time the benefits of early diagnosis and treatment are irretrievably lost, and they often end up paying

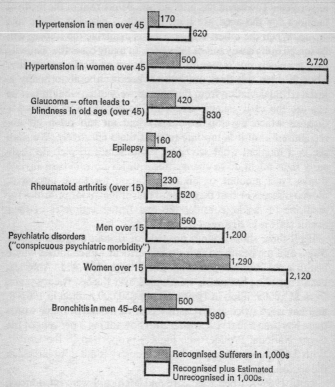

FIGURE 11.3  The clinical iceberg – general population

These figures are for the whole of England and Wales. They were extrapolated by the Office of Health Economics, London (in their publication 'New Frontiers in Health') from an original estimate based on an average general practice by J. M. Last ('The Iceberg', *Lancet* [1963], 2, 28–31).

crippling hospital bills. To add to their burden, because they are high-risk groups the poor usually have to pay higher premiums than the wealthy. Because they largely rely on fee-paying patients, hospitals and doctors have tended to mushroom in the richer areas, leaving great pockets of scarcity, so that many people who

FIGURE 11.4   The clinical iceberg – the over-65s

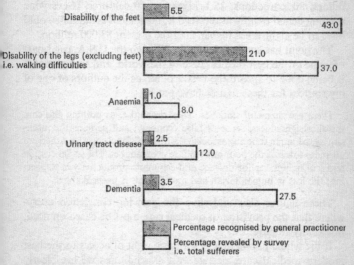

From *The Care of the Elderly in Scotland.* Royal College of Physicians (Edinburgh, 1963).

want and can even afford medical care are unable, or find it inconvenient, to get it.

What is this system with its spectacular overall expenditures doing for American health? Here are a few examples:

Compared with ten other developed countries, the U.S. is sixth in general mortality, tenth in the percentage of deaths under five years, and eleventh in infant mortality.

Nearly one in five young men is disqualified from military service for health reasons.

3·5 million poor children under five need medical help but do not get it under public-assistance programmes; more than a third of pre-school children who need eye treatment do not see a doctor; 3 million children who need glasses do not have them.

Forty-five per cent of children between 5 and 14 have never been to a dentist, although 97 per cent of all 6-year-olds have tooth decay. Every 100 young people joining the forces need 505 fillings, 80 extractions, 25 bridges and 20 dentures. If everyone needing dental treatment received it, the annual dental bill would have to be eight times higher than the present $3,000 million.

The point has surely been made. Though the U.S.A. can boast of some of the best medical care in the world, and though much is being done to spread the best around, as the authors of one of my sources for these figures have put it:

These are shameful statistics. They dramatise, as nothing else can, the shortsightedness, sense of false economy, and perhaps the insensitivity to human suffering associated with traditional negative American attitudes towards the poor and welfare programmes. The social cost, in terms of manpower for defence and economic productivity is staggering; the loss in humanitarian and moral terms is unspeakable.*

These figures also underline the growing conviction everywhere that the provision of medical care must be based on need, not ability to pay.

If the whole population has a basic right of access to the best modern medical care, despite vast expenditures we are clearly failing to satisfy that right – in the U.S.A., Britain and elsewhere. And with expectations rising, costs soaring, vast gaps in the provision of care being exposed, the future looks rather tough. Medicine *is* in crisis. Can we cure it?

Many people have a touching faith that miracles of medical science will pull us through. Anti-viral drugs will give the *coup de grâce* to all infections; there will be drugs to clear out our arteries and prevent heart disease; there will even be drugs to put an end

---

* Roger Battistella, Richard Southby, 'Crisis in American Medicine', *Lancet* 1 (1968), pp. 581–6. Other figures are from various publications by the U.S. Department of Health Education and Welfare.

to cancer. Perhaps there will; or at least there will be partial solutions to these frightful ills, and we shall be profoundly thankful for them. But will they get medicine out of its crisis? Few experts think so.

For one thing, there is no guarantee that they will cut the direct medical costs of their particular diseases significantly. For example, the Office of Health Economics has estimated that if *everyone* getting cancer were cured by a cheap drug regime, involving thirteen weeks' treatment at home at a cost of £5 a week, the National Health Service would save only £22·6 million a year – or less than 2 per cent of its budget. The reason is that because it usually kills rapidly, with little hope of stopping it, present treatment and care costs for cancer are not high. Similarly with those promised anti-viral drugs. Mass prevention campaigns against infectious diseases usually increase medical costs: for example, with poliomyelitis vaccines now costing as little as ten pence a shot, mass immunization is still more expensive than treating and caring for the few who used to get polio. Only with very common diseases does immunization pay off economically: streptomycin for tuberculosis is a classic example from the recent past, and influenza vaccines will be another in the near future. Of course, higher medical costs are only part of the picture: the total economic savings from the reduction of sickness and the extension of working lives – let alone the saving of grief and suffering – can be enormous.

Equally important, we can no longer believe, as we used to, that if there are N diseases and medicine finds cures for P of them one ends up with only N minus P to deal with. Disease is a process involving the adaptation of a whole society to its environment and has a fantastically more complex basic structure than this crude idea suggests.* More simply, when you sweep back one disease something else will take its place – possibly a few years later in the life cycle or (as the rising tide of young deaths from

* For a masterly account of this complexity see *Man Adapting* (Yale University Press, 1965) or the shorter, more popular *Man, Medicine and Environment* (Frederick A. Praeger Inc., New York and London, 1968), both by René Dubos.

car accidents shows) possibly not. Some of these new diseases might be easy to control, some of them might not be; but as we cannot predict what they will be this is perhaps something we can leave for our descendants to face.

In the meantime, how can we deal with our own health crisis? One apparently obvious answer is to spend much more on health, mainly to produce a vast increase in health manpower. About 70 per cent of health spending is on salaries, and, as scores of reports have said, the health crisis is largely a manpower crisis. The answer would be more cash, more attractive salaries, more medical schools, etc. Of course, it could be done; and when one considers that in Britain the total N.H.S. budget is roughly the same as the amount spent on tobacco, or on alcoholic drinks, or on gambling, there is a strong case for saying it ought to be done. (Alternatively, perhaps we just do not care enough.) Unfortunately, we also grossly underspend on social welfare, housing, education and just about every other publicly financed activity. Education, for example, is facing a manpower crisis just as serious as medicine's: there are now few countries which do not face a chronic shortage of primary and secondary school teachers. With all these other pressing priorities it is now generally accepted that medicine must stick within its limit of little more than 5 per cent of national resources.

So most medical administrators are turning to increasing productivity and efficiency. '*The crisis, however, is not simply one of numbers . . . Unless we improve the system* [original italics] through which health care is provided, care will continue to become less satisfactory, even though there are massive increases in cost and in numbers of health personnel,' states the U.S. manpower study mentioned earlier. Its message is now thought to apply everywhere.

Scores of ideas for boosting medical efficiency are in the air. With hospitals, a faster tempo would make a huge difference. For example, Dr Robert Logan of Manchester University has estimated that if the British national average of twelve-day stays in hospital were cut to the nine days some regions have achieved, hospital waiting lists could be cleared within a year. There would then be spare beds and skills for the chronic diseases and handicaps which urgently need a spell of specialized in-patient treatment. Automatic machines for laboratory tests could make all the difference here: instead of waiting two days, doctors could have test results back in two hours. Computers could also boost efficiency greatly, for example by handling patients' records, routing patients to hospital so that specialist units are not choked up here and under-used there, and eventually even helping with diagnoses. Standardized hospital designs and components could slash hospital building costs significantly: in Britain health authorities are pressing this theme hard. Further off some enthusiasts foresee giant three-thousand- to five-thousand-bed hospital villages as the heart of the hospital service. On the grounds that bigger is cheaper, these huge 'care campuses' would bring all the specialists – brain surgery, heart surgery, kidney machines, research and so on – under one set of roofs, with great savings in shared overheads, nursing, etc.

Other ideas aim at improving the family-doctor and community-health services. The need for efficient reorganization here is enormous. If every one of the 54 million men, women and children in Britain saw their G.P. for just one hour a year – perhaps for those much-recommended regular check-ups – each of the 23,000 G.P.s would have to be at it (and at nothing else) for 48 hours a week. Group practice is one major thrust here; comprehensive community-health centres providing everything from regular automated screening check-ups to simple health advice are another; a great increase in the use of medical social workers and other kinds of 'lay doctors' is a third. In Britain, in the interests of efficiency if not of patient convenience, there might also have to be a major swing away from home visits by doctors. Many British G.P.s spend half their working week on home visits and much of their time in surgery working on the level of a clerk or nurse. If they are to be streamlined they must stay in their surgeries and be paid enough to afford secretaries and nurses.*

* The British G.P. does more home visiting than any other doctor in the world. Scandinavian doctors, who come second in home visiting, do only a fifth as much. The usual Scandinavian practice is to take domiciliary

All these schemes may go a long way towards averting the looming health crisis, but they have a great amount of resistance to overcome and may take years to put into action. Even then they may not do the trick. Consequently, most experts are convinced that medicine will have to go much further. Instead of just trying to prolong life and cure sickness in any way at any expense, it will have to realize that it is an industry, and apply ruthless business-efficiency methods to its goals. It will have to start costing life, death and ill health carefully.

There is a good analogy here with the power industry. Some power stations produce electricity cheaply, others are inefficient and expensive. Long ago the power industry realized that it was madness to use these stations indiscriminately; instead they worked out ways of running the most efficient stations continuously to meet the 'base load' of power demand, only bringing in the most costly stations to meet peak demands. Similarly with medicine. If the health industry is to be made really efficient, the argument goes, it will have to find out which are the cheapest kinds of death or illness to prevent. Having found them it will have to pump resources into these cheaper strategies until they are fully backed up, and only then go on up through the more expensive ones until it has to stop when the money runs out. Only thus can it maximize its death- and sickness-preventing powers. But whether we can accept the human consequences of this cold-blooded rationalization is another matter.

This cost-effectiveness idea is certainly not new to medicine. To some extent individual doctors have always had to list their patients in an order of priority, because they do not have the time or facilities to deal with them all – especially in such emergency situations as might occur on a battlefield. And wherever governments control health spending they have to make some attempt

---

patients by ambulance to a health centre, do the relevant tests there, and send them home again before a doctor is brought in. This is said to be both more medically useful and economic. As for the U.S.A., it is often said that even a New York millionaire can hardly get his doctor to visit him at home.

to allocate resources between different branches of medicine. What is new is the scale and pace with which cost-effectiveness ideas are being pursued and built into health planning.

Since this looks like being the basis of most political discussions of health in future, let us look at length at some examples, starting with the emphasis on the prevention of death. Some fascinating, disturbing and myth-shattering ideas emerge.

To begin with, what about those high peaks of modern surgery – transplants and artificial organs? Are they an inordinately extravagant way of trying to avert death?

At present we can do reasonably accurate sums only for the kidney, because only with kidney transplants and machines do we know for how long on average they postpone death. But let us assume that in a few years heart transplants and the artificial heart do as well as the equivalent kidney operations in the best centres today. In other words, with any of the four procedures given below, roughly 90 per cent of patients live a year, 80 per cent two years, 70 per cent three years . . . until we get 50 per cent living five years and a similarly decreasing proportion living longer. On this assumption, because the long-term survivors balance out the early deaths, the mean survival is five years and the average cost of averting a death for this time equals the initial cost of the operation or machine plus the running costs for five years. So, bearing in mind that these figures are necessarily very rough, this is the result:

TABLE 11.1*

| Costs in pounds: | Initial cost (i.e. operation/ machine) | Running cost per year | Cost per death averted for 5 years |
|---|---|---|---|
| Kidney transplant | 1,000 | 200 | 2,000 |
| Heart transplant | 3,000 | 200 | 4,000 |
| Kidney machine (hospital) | 2,000 | 1,500 | 9,500 |
| Artificial heart | 4,000–16,000 | 200–2,000 | 5,000–26,000 |

*Most of the figures are based on those given in Chapters 9 and 10. The kidney and heart transplant operation costs are widely quoted ones (cf. £2,000 for average major heart surgery in Britain, including all preliminary tests and preparations plus intensive care afterwards). I have

These may be very rough figures, but they certainly point the difference between the cost-effectiveness of 'once-and-for-all' transplants and the machines with their continual running costs – a difference that of course gets larger if patients survive longer.

Now let us look at a range of other cost estimates. Because most estimates are based on rather different assumptions and can rarely be compared meaningfully, I have taken nearly all that follows from a series of studies published by the U.S. Department of Health, Education and Welfare.* The purpose of these studies was to look at several major 'diseases' and see which should get priority for new control programmes during 1968–72. For each programme and disease they were interested in two figures. The first was obtained by dividing the cost of the new programme by the number of lives it might save, to give the *cost of averting a death*. (The second figure, which measures the economic value of the programmes, we shall come to later.) Some of their answers are shown in Table 11.2.

Again, these are obviously not firm figures. A stop-smoking campaign, for example, could misfire badly or with brilliant advertising could catch on much better than expected. Yet already at least one crucially important fact has turned up: preventive medicine can be cheap or very expensive. Why is this? Why, for example, those huge variations in the cost of the cancer screening programmes?

The idea of cancer screening is to detect the early 'hidden' signs of the disease by some special test long before their owner

---

assumed £4 a week for drugs, check-ups, etc. to get the running costs.

For the artificial heart, the lower figure is for battery-powered models, the higher for isotope power. $10,000 is a widely quoted figure for implanting a full heart, with $30,000 on top of this for the isotopes. I have assumed the same running costs as for transplants – the hearts are supposed to work well after all. For battery models the extra cost of recharging would be negligible but for the isotopes there would be an additional $2,000 to $3,000 a year for repayment of interest, re-processing the isotopes, surgical fees for replacement when they are used up (see Chapter 10).

* Selected Disease Control Programmes: September 1966 (Department of Health, Education and Welfare, Washington, D.C.).

notices any symptoms. Once detected, the incipient growth can usually be removed, thus preventing a full-fledged, fatal cancer. But since the signs are hidden, everyone at risk must be screened

TABLE 11.2

| Programme | Deaths Expected 1968–72 | Deaths Averted 1968–72 | Cost per Death Averted $ |
|---|---|---|---|
| *Cancer prevention* | | | |
| Uterus & Cervix | 70,000 | 34,200 | 3,470 |
| by early detection and operation | | | |
| Lung | 250,000 | 7,000* | 6,400 |
| by 'stop-smoking' education and advertising | | | |
| Breast | — | 2,400 | 7,660 |
| by early detection, etc. | | | |
| Head & Neck | — | 270 | 29,100 |
| by early detection, etc. | | | |
| Colon & Rectum | — | 170 | 42,900 |
| *Syphilis prevention* | 15,500 | 15,070 | 22,900 |
| by extending present programme (currently 630,000 cases) | | | |
| *Tuberculosis prevention* | 40,000 | 7,200 | 28,000 |
| by extending present programme (currently 100,000 cases) | | | |
| *Arthritis control* | Very few | Very few | More than 100,000 |
| by building clinics to help early diagnosis and improve care (currently about 13,000,000 sufferers) | | | |

*Lives saved to 1972 only.

Note that Deaths Expected refers to deaths without the proposed new programmes. In all columns I have rounded off the numbers. Much of this study was concerned with analysing different programmes for each disease, but I have selected the most effective in each case.

and this usually involves vast numbers: e.g. with cervical cancer, ideally all women over twenty who have had sexual intercourse. So for a disease to be suitable there must be a simple, cheap test (so that vast numbers *can* be screened); the disease must be common (so that testing is productive); and there must be a good hope of cure once detected (so that screening has any point at all). To see the difference these three factors can make, here are some other American estimates for uterine and colon-rectum cancer, shown in Figure 11.5.

FIGURE 11.5

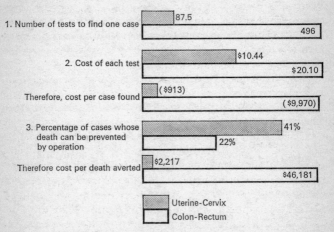

1. Number of tests to find one case — 87.5 / 496

2. Cost of each test — $10.44 / $20.10

Therefore, cost per case found — ($913) / ($9,970)

3. Percentage of cases whose death can be prevented by operation — 41% / 22%

Therefore cost per death averted — $2,217 / $46,181

☐ Uterine-Cervix
☐ Colon-Rectum

From 'Cost-Benefit Analysis and Social Planning', lecture by Robert N. Grosse, U.S. Department of Health, Education and Welfare.

For the future of screening, the most critical factor is the cost per test (as well as the education of the public to go for screening when it is available). This is not just because cheaper tests – those that save manpower – make screening more economically justifiable. Unless costs are very low, whatever the gain in cheap life-saving the cost of screening everybody may still be prohibitive (a bun may be cheap at a penny, but who can afford a million buns?). Also, even if the country could afford this, the cost of screening everybody could still be much higher than the conventional approach of letting these diseases take hold and trying to cure the few advanced cases that one can. As a result, screening is still rationed: in Britain, for example, N.H.S. cervical tests are offered only to the over-thirty-fives (though younger women who apply are not turned away).

This is bound to change because there is another reason why the cost per test is critical: it is the only one of the three factors that *can* be changed dramatically. New automatic machines are coming on the market that can do cervical cancer tests five times faster and at less than one third the cost than before. With them, it has been estimated, all the 15 million women in Britain who are over 20 and are at risk could be screened yearly using only 400 machines and 400 cytologists at a cost well under fifty pence a head.*

When this happens and everyone is screened regularly from early years, the third factor – the cure rate for the disease once it is detected – should also improve. A good example of how this happens comes from lung cancer, the commonest type of cancer in Britain and one of the most frequently fatal (each year it kills nearly 30,000 men and women, causing 40 per cent of all male cancer deaths). Apart from anti-smoking campaigns, the main prevention hope is the mobile mass X-ray unit. A recent study† has shown that with early detection by X-ray the proportion of cases where removing a lung can still give a chance of life goes up from 33 to 47 per cent and for those who are operated on the two-year survival-rate goes up from 37 to 46 per cent. In other words, for every 100 cases picked up, 22 rather than 12 will live an extra two years. With X-rays costing only £30 per 100 these lives are bought very cheaply. If all the 700,000 men in Britain over 55 who smoked fifteen or more cigarettes a day were X-rayed twice a year, the cost would be £420,000, but it would mean that 2,400 rather than 1,400 lived another five years. So

---

*This example is for a machine produced by Vickers Instruments Limited, first tested on a large scale in 1968. Cervical cell scrapings, one from each woman, are automatically fixed and stained on a transparent tape and fed through a microscope. A computerized scanning system attached to the microscope 'recognizes' any larger than normal cells and automatically punches the tape at that point. When the tape is run back through the microscope it stops at that point and the cytologist examines it by eye, thus saving him searching every specimen from every woman.

† *British Medical Journal*, 2 (1968), pp. 710–11.

each of those years would be bought for only £84 worth of screening (plus operating costs, etc.).

This raises the interesting problem of whether people who deliberately risk their own lives should get preferential treatment for preventive medicine. The cost of picking up potential lung cancer survivors by screening *non-smokers* over 55 is just over five times higher than with people of the same age group who smoke 25 cigarettes or more a day. Society would do five times better if it screened the smokers only and let the well-behaved look after themselves.

As costs drop, the number of diseases for which there are cheap, clinically meaningful tests will almost certainly increase. The essence of the chronic and degenerative diseases – from anaemia, rheumatism and diabetes, to cancer and the cardio-vascular diseases – is that they develop slowly and so must produce some biochemical or cellular 'fingerprint' that can be detected. The great difficulty is sorting out the genuine 'finger-prints' which indicate presymptomatic disease from normal variations between individuals. But we can expect rapid progress here, so that many useful predictive tests will be developed, though with a wide range of cost and eventual 'curability'. As a result, the cost-effectiveness of screening will change and each disease programme will have to be constantly reviewed. Today's 'best buys' may become relatively bad bargains tomorrow. This could create some difficulties. For example, how fast does one push a screening programme that is already well established when it is overtaken in the 'best buy' list by new diseases? And how urgently does one push a new ultra-cheap screening possibility that occurs? As an example of the latter, in February 1968 a group of Leeds doctors reported that a low level of the hormone oestriol in a pregnant woman's urine is a reliable indicator that her child might be stillborn, while early delivery improves the chances that it will live. If widely used, the cost per death averted would be a mere £50.*

Now we come to one of the most difficult of all questions in

---

* R. A. Heys, *et al.*, *Lancet*, 1 (1968), pp. 328–33.

cost-effectiveness studies. In the harsh world of medical crises it is no longer considered enough to define a 'best buy' solely on the cost of preventing death. One must also consider the total economic benefits of any programme. Some diseases mainly kill young people with a lifetime of earning capacity ahead of them; some hit the unproductive old. Clearly it is far more economically productive to prevent the first kind than the second. Other diseases are mainly cripplers rather than killers. Preventing them may not save many lives but can vastly increase the earning capacity of the people one saves and slash their continual medical and social welfare costs. So the medical economists have started working out the investment potential of medical-care programmes. What they do is to add up what society can expect to gain (from extra earnings, reduced costs, etc.) and divide this by the cost of the programme to get a benefit-cost ratio.

The benefit-cost ratio can make a huge difference to which programmes are 'best buys', as this new version of Table 11.2 shows:

TABLE 11.3

| Programme | Deaths Expected 1968–72 | Deaths Averted 1968–72 | Cost per Death Averted $ | Benefit-Cost ratio |
|---|---|---|---|---|
| Cancer prevention | | | | |
| Uterus & Cervix | 70,000 | 34,200 | 3,470 | 9 |
| Lung | 250,000 | 7,000 | 6,400 | 6 |
| Breast | — | 2,400 | 7,660 | 4·5 |
| Head & Neck | — | 270 | 29,100 | 1 |
| Colon & Rectum | — | 170 | 42,900 | 0·5 |
| Syphilis prevention | 15,500 | 15,070 | 22,900 | 12 |
| Tuberculosis prevention | 40,000 | 7,200 | 28,000 | 8 |
| Arthritis control | Very few | Very few | More than 100,000 | 42·5 |

Though the first three programmes would save the most lives for the least money, the three with the largest economic pay-offs are cervical/uterine cancer, syphilis and arthritis. And it was these three that the Department of Health, Education and Welfare recommended ought to get the money before any of the other programmes were initiated or expanded.

Benefit-cost studies are fraught with errors, simply because adding up the expected income of the 'saved' is such a crude measure of their economic value. It also ignores the fact that whatever we earn we all cost society a great deal in hidden ways. As some wags have put it neatly, what do you do in a country like Britain during the 1960s with a chronic balance of payment *deficit* where an obvious way of clearing the deficit would be to kill everybody off? More seriously, it ignores factors like a person's productivity after he is 'saved'. For example, a recent British study shows that if instead of treating disabled people as unemployable and paying them pensions not to work they were re-trained and encouraged to go back to work (possibly in special sheltered workshops) the country could be saved huge sums.*

But this assumes that they would use capital resources as efficiently as normal people, with the same productivity, and would not lower the productivity of their normal fellow workers. This is not to argue that rehabilitation should not have a high priority on humanitarian grounds; only that those very high benefit-cost figures for disabling diseases may be more than a little inflated. If we are to have our lives and illnesses cost-benefited, there is a pressing need for more careful estimates.

Now for just a few more figures. Consider Table 11.4, for example, which shows how badly medicine proper can score compared to other ways of averting death or disability. They also come from that U.S. Department of Health, Education and Welfare study.

Again, the estimates cannot be exact. Some programmes could misfire badly, others could exceed even the most optimistic expectations. As an example of the latter, when Britain intro-

---

*For instance, a 44-year-old man with a leg amputated at the hip, instead of costing £14,000 during his lifetime in unemployment benefits, etc., would earn £20,000 if he went back to work at the average national wage – boosting his own earnings by £420 a year and saving the government £580 a year. J. McKenzie, Office of Health Economics (London), to International Congress on Rehabilitation, 2 July 1968.

TABLE 11.4

| Road death prevention programmes (250,000 deaths plus 20 million casualties expected in 1968–72 from all kinds of road accidents) | Deaths Averted 1968–72 | Cost per Death Averted $ | Benefit-Cost Ratio |
|---|---|---|---|
| Use your seat-belt campaign | 22,900 | 87 | 1,351 |
| Campaign to teach pedestrians to cross roads safely | 1,650 | 670 | 144 |
| Buy and wear motorcycle helmets campaign | 2,400 | 3,340 | 56 |
| Don't drink and drive campaign | 5,300 | 5,820 | 22 |
| Driving licences only after medical check-ups | 400 | 13,800 | 4 |
| Nation-wide campaign to improve driver skills | 8,500 | 88,000 | 2 |

duced breath tests and legislation against drinking drivers in October 1967, road deaths the following month dropped by 20 per cent and serious accidents by 15 per cent on the previous November. A year later, road deaths were down by 730 (7 per cent) and casualties by 6 per cent on the previous year. Much of this reduction was ascribed to the breath tests. Though no one seems to have costed this programme it is likely to prove a cheap way of buying lives: the breathalyser instrument itself costs only twenty pence or so while only 3,000-odd tests were done in October to launch the November drop.

The lesson of these figures is surely that compulsion and the fear of being caught can often buy lives far more cheaply than persuasion or medicine; and that if we care about these lives we will either have to accept these restrictions or live much more sensibly. Actually, there is a wide spectrum here in the extent to which we might mind restrictions. At one end there are restrictions that nearly everyone approves. For example, in Britain recently manufacturers of children's clothes were forced to flameproof them (at an estimated cost of twopence a garment) thus saving countless lives and injuries for decades to come at negligible cost. How many posters or lectures on safety in the home would be needed to achieve the same results? Many similar rules have been or could be imposed – from the present stringent public-health regulations on food and sanitary standards to, say, the compulsory fitting of anti-tip saucepan guards on cooking stoves. In other cases many people start by resenting the restrictions but soon accept them when they see that they save lives: the rule that you only drive on the left or right is an obvious one from the past, the compulsory *wearing* of seat-belts (perhaps with their buckles wired to the ignition circuit so that we cannot cheat) is a likely one for the future. However, one must watch costs. Since September 1970 all farm tractors in Britain must have a canopy to protect the driver in case the tractor tips over. This should save thirty to forty lives a year at a total cost of £4,000,000: a phenomenal £100,000 or more per death averted.

Much more difficult to accept – and therefore for a government to introduce – are laws which do not visibly save lives for decades. This time lag is the great legislative and educational stumbling block for so much preventive medicine. Anti-smoking legislation may be years away, and anti-smoking persuasion campaigns have little impact, simply because lung cancer in thirty years' time seems such a ludicrously improbable notion, whatever the statistics say. If the chances of dying as a result of smoking were the same as now, but death followed in months rather than decades, there would be no need for legislation or education. It is inconceivable that cigarettes would be sold at all.

Lastly, just one more set of figures – this time on programmes that have little to do with preventing death but a great deal to do with the enhancement of life. In another series of studies the U.S. Department of Health, Education and Welfare asked itself what disabilities it could prevent for each $10 million per year it spent on various programmes in health depressed areas. Table 11.5 shows what it found.

So preventing these defects, while not all that cheap, is cheaper than any of the death-averting programmes we have looked at – except for the non-medical ones. And as for that adult education programme, it gives a better benefit-cost ratio than most of the medical programmes. Resource allocation is not a straightforward

TABLE 11.5

*Reduce physical defects by*
*'seek out and treat' programmes:*

| | | |
|---|---|---|
| Physical handicaps prevented or corrected | 1,870 ⎫ | |
| Vision problems prevented or corrected | 3,470 ⎬ | $790 per case |
| Hearing loss prevented or corrected | 7,290 ⎭ | |

*Reduce number of 18-year-olds*
*with decayed and unfilled teeth:*

| | | |
|---|---|---|
| Comprehensive dental care without fluoridation | 18,000 | $550 per case |
| Fluoridation alone | 294,000 | $35 per case |

*Basic education for 30,000 adults*

| | | |
|---|---|---|
| Increase their total earnings by | $110 million | Benefit-Cost Ratio of 11 |

business! To sum it up – more for fun than anything else – in Table 11.6 I have put together for comparison several of the figures discussed so far.

TABLE 11.6   What can £1,000,000 buy?

| | |
|---|---|
| Crash canopies on all U.K. tractors | 10 lives |
| Isotope-power heart: 5 years of life | 38 lives |
| Artificial kidney:   ,, ,, ,, ,, | 105 lives |
| Syphilis prevention | 110 lives |
| Heart transplant: 5 years of life | 250 lives |
| Breast cancer screening | 330 lives |
| Cervical cancer screening | 720 lives |
| Lung cancer X-rays for old smokers | 2,400 lives |
| Prevent or correct physical defects | 3,160 — |
| Pedestrian injury campaign | 3,750 lives |
| Oestriol screening in late pregnancy | 20,000 lives |
| Seat-belt campaign | 29,000 lives |
| Fewer decayed teeth by fluoridation | 73,000 — |

Of course, some of these figures are not realistic ; e.g. there are not 29,000 road deaths to be saved by a £1,000,000 programme on seat belt use.

These figures seem to highlight many of the main tensions in modern medicine – tensions that will increasingly occupy us in future. So for the rest of this chapter I propose to try to outline them.

First of all, most medical specialists are angry (or just amused) by attempts to cost their own particular corner of medicine. If one is doing one's best to develop a new treatment such as the kidney machine by making it cheaper, more effective, more curative, more life-saving – and is using it meanwhile to save lives – little men with slide rules who say that you are being extraordinarily extravagant with resources are hardly going to be welcomed with open arms. More generally, doctors have always cared for their patients individually and on the basis of what is technically possible. If a common cold cure was possible they would give it to you; now that deep brain surgery *is* possible, if you need it they will give it to you as best they can, within the limits of their skills and resources. But what if someone said that both were too expensive? Does this mean that cost-effectiveness cuts their traditional freedoms, replacing the doctor's judgement and conscience by the invincible slide rule?

I do not think so. The point about cost-effectiveness is that it is only a method for guiding *future* developments, not for banning present established procedures; or if it does restrict the latter then no one would mind much. No one is suggesting that it should be used to tell doctors not to investigate and treat a critically ill patient because it would cost too much. No one is talking about running down existing programmes, like kidney machines, on the same grounds. No one is talking about halting research into this or that. All cost-effectiveness is about is deciding where to *expand* our future horizons for care and research. Do we push this research frontier, or that? Do we expand kidney machines so that everyone in need can have them, or the more cost-effective transplants? Or do we slow down the growth of both these procedures in favour, say, of cancer screening? It is concerned with what new *advantages* we should seize, where we should put tomorrow's men and resources, not with taking away what we have now.

Nevertheless, it will not be easy. We already face the nasty situation that not being able to get a kidney machine exacerbates the misery of dying of kidney failure. Similarly, with other advances, people will go on having to say, 'I could be saved if only ...' and the public will go on crying 'Scandal'. It would all be so much easier if every new technique could be costed the moment it leaped into its inventor's mind or bubbled up in the test-tube; then if it was inordinately expensive it could be quietly suppressed before anyone else knew what they might be missing. But unfortunately one cannot cost anything until it has been developed and tried and perhaps saved lives and, above all, been headlined as another miracle of modern medicine. So once again perhaps the most pressing need is for public discussion and awareness of the problems. Families accept that they cannot buy everything they want; we must realize that the same is true of our lives and health as a total community.

The second main problem is the tension between prolonging life and treating ill health or disability. Is medicine for averting death at all costs or, at the other extreme, for coping with all the minor and not so minor ailments that assault the vast majority who are not threatened by death yet? Transplants or psychiatry? Kidney machines or contraception? Of course, it is for the whole range, but very broadly where should the emphasis be? Do we have our priorities right?

The trouble here is that there is absolutely no logical way of choosing except by some measurable quality like benefit-cost ratios. One cannot make intuitive judgements that it is better or worse to spend £X on prolonging one life for a year than to spend it, say, on alleviating the misery of arthritis in fifty people. Yet though medicine has traditionally put its greatest efforts into life-prolonging, what those benefit-cost figures show is that for the health and wealth of society less dramatic goals can often be a relatively far better bargain.

Again, we have to remember that we are talking about the future, not about the present; and that even with life-prolonging there is a vast range in the degree to which we would mind being 'rejected'. To make this clearer, consider four examples. First, a doctor stands over one's child and says, 'Sorry, he's got meningitis but although we can save him we're not going to because of the cost.' Second, a doctor tells a healthy woman that he cannot give her a cervical cancer test because he thinks there is little point in them considering the trouble involved. Third, a government does not bring in anti-smoking or seat-belt legislation this year. Fourth, a government announces that education (but not health) spending will go up by 1 per cent, a decision that in the last analysis says it is wrong that children should be educated by one-thirtieth of a person while some people are kept alive by the full-time devotion of one person. Everyone would be horrified by the first; few would really mind the second; and no one would care much at all about all the lost lives that the last two decisions entail.

The death/ill-health tension underlies the well-known conflict between the hospitals and the general medical services; the medical specialists and the family doctors. In the last two decades most advanced countries have invested more heavily in their hospital services than in general medicine. In Britain, for example, in the last fifteen years the proportion of the National Income spent on the hospital service has risen to about 125 per cent and the share for general medicine has dropped to about 75 per cent of the 1950 figures. While this has happened for many complex reasons, a major cause has been the way the public has been sold on those glamorous, shiny, 'scientific' gadgets for keeping us going: 'To secure vast public support for research and for the latest equipment and techniques for saving or maintaining life is relatively easy. On the other hand, the organizers of medical care, on which the effective application of scientific knowledge depends, have a hard row to hoe.' ( Gordon McLachlan, Secretary of the Nuffield Provincial Hospitals Trust.) *

Of course, general hospitals are a vital part of medical care, but they are to a large extent a service to the acutely ill, to those whom death may threaten. They hardly touch what one doctor has called the major causes of ill health in advanced countries today – 'a confusing mixture of disease, maladaptation, faulty relationships, poverty, poor education, ignorance, obstinacy, fear, virtue and vice'.† No specialist or hollow-eyed houseman in the hospital can peer into this tangled undergrowth of ill health. It can only be penetrated by a network of local services based on the

* Gordon McLachlan, 'Medical Science to Medical Care', Lancet, 1 (1967), pp. 629–31.

† Dr E. B. Thornton in World Medicine, 18 July 1967.

family doctor and community health centres – a network which forms the *base* of the health–care pyramid and only passes on the minority of cases (seldom more than 10 per cent) that require special tests or acute treatment to the hospitals.

We have to back this grass-roots kind of medicine heavily in future, for many reasons. Above all, perhaps, because thanks to the triumphs of scientific medicine, so many people are old; and in the very near future so many more will be old. Yet it is on this fact that the benefit-cost approach to medicine founders. In Britain at the moment about 1,000 people pass retirement age every day, many of them facing another quarter century of life. These people are no use at all in the cold calculations of the benefit-cost accountants. They are caught in the dilemma that just as they are the logical outcome of successful scientific medicine, so they are the logical victims of the economic choices that scientific medicine is forcing on us.

And there are other victims. Besides the old, perhaps the group for which medicine and society should blush most deeply are the mentally ill. In Britain at least one in four of all consultations with family doctors is about emotional- or mental-health problems – problems which most of these doctors must face without training. More appalling, nearly half Britain's hospital beds are occupied by psychiatric patients. Nearly half the beds, but nowhere near half the services. On average there is one consultant for every 150 mentally-ill patients, and they get only 15 per cent of total National Health Service funds and a tiny fraction of medical research funds.

Faced with these people the medical economist's approach must fill us with anxiety unless it is mixed with a great deal of humanity. Without humanity, as the eminent Swedish doctor Gunnar Biörck has written,* the future of cost-effective medicine may look like this:

We are likely to get two privileged groups, and one neglected group, among our patients. The healthy wage-earner will become the privileged, for whom a battery of health-screening procedures will be available without cost. Privileged also will become anyone suffering

* *British Medical Journal*, 2 (1965), p. 7.

from a sufficiently interesting disease to warrant special investigations
and the assemblage of technical experts for diagnosis and treatment.
The underprivileged will be the aged, the worn out, the deteriorated and
perhaps, still more, the psychologically maladapted – in short, the
useless, the uninteresting and the nuisance.

Useless to the economist; uninteresting to most hospital
doctors; a nuisance, often, to their families, these are the people
for whom community services – from family doctor through
medical social worker to social clubs for the lonely – must be
expanded to serve. At the moment they are virtually ignored.
These people are perhaps the supreme example of the way modern
scientific medicine, with all its emphasis on acute physical illness
and the prolongation of life, is by-passing the major health needs
of the majority of the population.

We all have to die some time and we shall all be grateful to
scientific medicine when it is our turn to have our deaths postponed (unless they strive too officiously to keep us alive). But in
the meantime we have to live, and it may be that the quality of
our lives should carry more weight. Perhaps medicine should help
us live more fully and spend less in averting our deaths. Medicine
in crisis cannot do both to the full. In the end this is the biggest
choice that we have to face.

# INDEX

Abortion, 18–23, 37–51; *see also* Birth-control
  aborted foetuses for experiment, 164–71
  Churches and, 37–8
  clinics, 46
  consequences of refusal, 39–40, 43
  cost, 45n
  do-it-yourself, 51
  effect of liberal laws, 21, 44–9
  illegal, 19, 41–2, 44
  incidence of, 19–21, 41, 45n, 47–8, 49
  legality, 39, 45n, 47–8, 170–71
  medical attitudes, 38–9, 44–5
  methods, 41–2, 46–7, 171
  pills, 49, 50, 51
  psychological effects, 42–3
  safety of, 41–2
  sex choice and, 106
  social effects, 43–4
Abortion Act (1967), 20–21, 45
Achondroplasia, 129, 140
Aggression, control of, 236–7
Albinos, 130, 159
American Headstart Programme, 223
Amniocentesis, 168–9, 172
Amnioscopy, 167
Amphetamines, 220, 235
Anencephaly, 182
Ante-natal care, 166–7, 181–6
Anti-lymphocyte serum (A.L.S.), 291–3
Arginine, 158

Artificial insemination:
  adoption and, 86–7
  anonymity, 87
  by donor (A.I.D.), 86–93
  by husband (A.I.H.), 83–5, 108
  Churches and, 84, 89–90
  clonal reproduction, 110–15
  costs, 88
  donors, 87–8, 94
  egg banks, 102
  embryo banks, 102
  gonad grafts, 98–9, 104–5
  host mothers, 102–4
  hybrids, man-animal, 115–16
  incidence (A.I.D.), 86
  legal problems, 84, 90–91
  medical attitudes, 84–5, 88–9
  psychological problems, 91–3
  sex choice, 105–10, 113–15
  sperm banks, 83, 93–8
  technique, 82, 87–8
  test-tube fertilization, 99–102
Artificial organs, 246–80; *see also* Kidney machines; Heart, artificial
Artificial womb, 102, 177–8
Autistic children, 243–4
Azathioprine, 288
Azoospermia, 86

Barbiturates, 235
Barnard, Dr Christian, 297, 305
Barnes, Professor Allan C., 182–4
Behaviour control, 241–5; *see also* Brains

Behrman, S. J., 92
Belzer, Dr F. O., 322
Benzedrine, 235
Bereiter, Dr Carl, 244
Birth-control, 18–51; see also Abortion, Contraception, Population
  advertising, 27
  Catholic Church and, 23–4, 44n
  classification of methods, 18–23
  clinics, 22, 27
  domiciliary programmes, 26–7
  effect on population size, 35, 53–62, 68
  family size and spacing, 56, 61–71, 107–8
  ignorance of, 19–25
  problem families and, 26
Birth defects, 187–214; see also Genetics, Eugenics
  attitudes to, 205 ff
  biochemical medicine for, 202–5
  cystic fibrosis, 194–6
  defect 'engineering' for, 201–14
  incidence, 126–8, 162
  infanticide for, 206 ff
  phenylketonuria, see Phenylketonuria
  prematurity and dysmaturity, 182, 189–91
  provision for sufferers from, 211–14
  pyloric stenosis, 188–9
  risks of, 128, 130–31, 133, 134–5, 136
  spina bifida, 196–201, 202, 207, 209
Birth-rates, 56, 59–60, 62, 67
Births, unwanted, 65–6
Blaiberg, Philip, 298
Brain and head transplants, 301–2
Brains, 215–45
  behaviour control, 241–5
  early development of, 176–7
  electrode stimulation of (E.S.B.), 227–33
  enhancing performance of, 174–7, 217–27
  memory, 217–20, 226–7
  'mood pills', 233–9
  overall structure, 216
Brewer, Herbert, 122
British Royal College of Midwives, 183
Bronchitis, 317
Butterfield, Professor W., 332

Callaghan, Dr John, 178
Calne, Professor Roy, 312
Cameron, Dr Ewen, 221, 225
Canada, contraception in, 22
Cancer, 292, 342–7
  cervical, 343–5
  lung, 317, 343, 345–6, 347
  prevention figures, 347
  screening, 342–7
Carrel, Alexis, 298n
Carter, Dr C. O., 189
Cartwright, Dr Ann, 65
Catholic Church:
  abortion and, 38–9, 44n
  artificial insemination and, 84
  contraception and, 23–4, 37
Cell hybridization, 111, 156–7, 158–9
Chemical warfare, 233, 237
Chimeras, see Hybrids
Chromosomes, 132
  defects, 132–5
Church of England:
  abortion and, 38
  artificial insemination and, 89
Clarke, Professor A. C., 173
'Clinical iceberg', 330–35
Clomiphene, 82
Clonal reproduction, 110–15
Colombia, birth-control in, 19
Colour blindness, 132
Congenital defects, see Birth defects
Consent to treatment, 270–71

Contraception, 18–37; *see also* Birth-control
diaphragm, 23
immunology and, 33
intra-uterine devices (I.U.D.s), 33–4, 74
oral, female, 22, 28–32
oral, male, 33
post-coital, 35–7
safe-period, 23n
sterilization, 33–5, 72, 96
sterilizers, reversible, 33–5
use, 28–30, 65
Contraceptives, research and development, 32–3
Cooley's anaemia, 142–3, 149
Cost-effectiveness studies, 339–52
Costs, 27, 45n, 83, 88, 149, 256–8, 262, 275, 280, 326–8, 333, 335–7, 339–51
Cylert, 221
Cystic fibrosis, 142, 149, 194–6
Cystomegalic virus, 182
Czechoslovakia:
abortion in, 41, 48
experiments on aborted foetuses in, 165

Darvall, Denise, 297
Davis, Kingsley, 70–71, 72, 73
Death:
attitudes to, 303–4, 315
definition of, 309–12
rates, 59, 60, 329, 331
DeBakey, Dr Michael, 267
Delgado, Professor José, 228, 233
Delinquency, and chromosome defects, 134
Dempster, W. J., 312–13
De Wardener, Dr H. E., 252, 258, 259
Diabetes, 202, 203
Dialysis, kidney, *see* Kidney machine
Diaphragm, contraceptive, 23
Dickinson, Robert, 86

Disease pattern, changes in, 327–8
Djerassi, Professor Carl, 32
DNA molecules, 128, 153–6, 219n
Dominant genetic defects, 128–30, 139–40
'Double' males, 134
Drinan, Father Robert E., 39
Drugs:
anti-depressants, 235
as brain performance enhancers, 219–27
drug-taking explosion, 238–41
hallucinogenic, 234–5
legislation on, 32
'mood' pills, 233–9
stimulants, 235, 236
tranquillizers, 235, 237, 238
Dunedin, Lord, 90
Dysmaturity, 182

Edwards, Dr Robert, 100, 101, 105
Egg banks, 102, 123–5
Egg grafts, 98–9
Embryatrics, *see* Foetal medicine
Embryo banks, 102
Engelmann, Siegfried, 244
Enzymes, 155–6, 158–9, 202–3
Epilepsy, 157, 229
Epiloia, 129
Erikson, Erik, 39
E.S.B. (electrical stimulation of brain), 227–33
Eugenics; *see also* Genetics
consanguinity bans, 148
economic consideration, 149–50
effectiveness of state programmes, 138 ff
genetic hazards of, 150–52
methods of applying:
genetic counselling, 135–9
screening for carriers, 141 ff
sterilization, 137, 140–41
negative, 126–60
scope for, 127
social considerations, 146–8

Eugenics:
  by marriage, 119–21
  germinal choice, 97, 122–5
  positive, 117–25
  selective breeding, consequences of, 119–20
  unintended effects, 121
Eutelegenesis, 122

Fallopian tubes, 99–100
Family, role of, 14, 74
Family planning, see Birth-control
Family Planning Act (1967), 20
Family Planning Association, 22, 28
Filshie, Marcus, 50
Foetal medicine, 161–86
  amniocentesis, 168–9, 172
  amnioscopy, 167–8
  ante-natal care, 166–7, 181–6
  artificial womb, 177–8
  brain performance and, 174–7
  decompression technique, 174–5
  rhesus-exchange transfusion, 172–3
  tests on aborted foetuses, 162–5
Foetal surgery, 173–4
Fox, Sir Theodore, 28
France:
  heart pacemakers in, 274
  transplant laws in, 314n
Fraser Roberts, Dr J. A., 136
F.S.H. (follicle-stimulating hormone), 82

Galactosaemia, 144, 145
Galton, Francis, 117
Gardner, Dr R. L., 105
Genes:
  description, 128, 153–4
  missing, 159
  synthesis, 155
  transplants, 155–7
  viral surgery, 158
Genetic counselling, 136–7
Genetic engineering, 153–60

Genetics; see also Eugenics, Birth defects
  carrier detection, 141–7
  counselling clinics, 136–7
  dangers from sperm banks, 94–5
  defects, incidence of, 127 ff
  genes, 128, 153–4, 155–9
  genetic engineering, 153–60
  heterozygote advantage, 143
  mutations, 139–40, 150–51
  risks of defect accepted, 136–7
  types of inheritance and defects:
    chromosome, 133–5
    dominant, 128–9, 139–40
    multi-factorial, 118
    recessive, 127–8, 130–32, 141–6
    sex-linked, 132–4, 141, 168–9
Georgetown University Hospital, Washington, 254–5
German measles (rubella), 135, 169
Germinal choice, 122–5
Glasky, A. J., 221
Glass, Professor D., 63
Glycogen storage disease, 144
Goitrous cretinism, 144, 203
Gonad grafts, 98–9, 104–5
Great Britain:
  abortion and, 21, 39, 41, 45, 46, 47, 48–9, 50
  artificial insemination and, 88–90
  birth-control legislation in, 20–2, 29
  birth-rates, 56, 59, 61–4
  contraceptive pattern, 22, 27
  death-rates, 59, 62, 328–9, 331
  doctors' visits in, 339
  excess of bachelors in, 108–9
  family size in, 61–2, 65–6, 69, 107–8
  genetic engineering in, 156–7
  health costs, 326–7, 330, 336–55
  heart pacemakers in, 274
  marriage age in, 62, 63
  mongolism in, 187–8

Great Britain – *Contd.*
  population growth, 53–4, 56–8, 59–63, 68
  pregnant brides percentage, 66
  provision for mentally subnormal children in, 211–14
  transplants in, 287, 292, 314–15, 316, 320
  unwanted pregnancies in, 63–6
  use of kidney machine in, 251, 253, 256–63
*Guardian*, the, 78
Gurdon, J. B., 110–11
Guthrie, Charles, 285
Guthrie test, 193, 207
Guttmacher, Alan, 49

Haemophilia, 110, 131, 141, 143–4, 159, 203
Haldane, J. B. S., 112–13, 122, 159, 185
Haller, Dr J. A., Jr, 173
Hallucinogenic drugs, 234–5
Hamburger, Professor J., 307
Hardin, Garrett, 35, 51
Hardy, Dr J. D., 295, 296, 297
Harris, Professor Henry, 156–7
Head transplants, 301
Heart, artificial, 264–83
  costs, 280, 341
  demand for, 277–9
  development stages towards, 265–75
  heart-lung implants, 269
  isotope powered, 274–5
  pacemakers, 271–2, 274, 281
  systems for driving, 271–5
Heart, description of the, 264
Heart transplants, 287, 297–300, 315, 316
  animal, 298n
  banking, 299
  cancer and, 292
  costs, 341
  demand for, 277–9, 316

  earliest, 297
  plus lungs, 301
  selection of patients, 299–300
  survival after, 298–9, 300
Hepatitis:
  mongolism and, 135
  serum, 262
Herrick, Richard, 286, 290
Heterografts, 323–5; *see also* Transplants
Heyns, Professor O. S., 174
Histidinemia, 144
Hitchcock, Dr Claude, 323
Hospital for Sick Children, London, 194n
Host mothers, 102–4
Hsu, Dr Y.-C., 179n
Human Tissue Act (1961), 314
Hume, Dr David, 285–6
Hungary, abortion in, 41, 48
Huntington's chorea, 140
Huxley, Aldous, 178, 234
Huxley, Sir Julian, 122
Hybrids, man-animal, 115–16; *see also* cell hybridization
Hydrocephalus, 197–8

Illegitimate births, 20, 63, 64, 74
Imuran, 288, 290
Incest, 94
Infanticide, 19–20, 174, 198, 206–11
Infant mortality, 185, 189–91
Infertility, *see* Subfertility
Inosinic acid pyrophosphorylase (IAP), 156
Intelligence Quotient (I.Q.), 118–21, 124–5, 174–6, 223–4
Intra-uterine devices (I.U.D.s), 33–4, 74
Israel:
  family structures in, 73
  sex ratio in, 109
Italy:
  abortion rate in, 23
  Cooley's anaemia in, 143, 149

Jacobson, Allan, 218
Japan:
    abortion in, 41
    first sperm bank in, 94
Jewish Orthodox Church, and arti-
    ficial insemination, 84

Kantrowitz, Dr Adrian, 267
Karim, Sultan, 50
Kerr, Dr David, 256, 258, 261
Keynes, Maynard, 58
Khorana, Dr H. G., 155
Kidney machine, 248–64
    costs, 253, 256–62, 341
    definition, 248
    drawbacks, 248–51, 262
    financial stress for patients, 253–4
    home use, 262–4
    is life worth living on, 252–3
    possibility of implanted, 264
    psychological stress for patients,
        254–5
    selection of patients for, 258–62
    survival on, 251–2
Kidney transplants, 285–93; see also
    Transplants
    banking, 299, 322
    costs, 341
    first, 285
    rejection, 288–93
    supply and demand, 316, 317, 318
    survivals, 288
Kleegman, Dr Sophia, 91
Klinefelter males, 133
Krech, Professor David, 215, 223

Lancet, 212
Lashley, scientist, 217
Laws and legal problems, 20, 21,
    23n, 32, 38, 39, 44n, 45n, 47–8n,
    72–5, 84, 90–91, 110, 137–8,
    165–6, 170, 171–2, 192, 206,
    314
Lederberg, Professor Joshua, 112,
    113, 115, 119, 176, 201

Leiden, Holland, Eurotransplant
    centre in, 320
Lesch-Nyan syndrome, 156
Liddicoat, Dr Renée, 175
Liley, Dr William, 172
Lillehei, Dr C. W., 269
Liver transplants, 294–5, 316n, 317;
    see also Transplants
Logan, Dr Robert, 338
Longmore, Donald, 301, 316, 320
Lower, R. R., 298n
L.S.D., 233–4, 238, 240
Lung cancer, 317
    smoking and, 342, 343, 345–6, 350
Lung transplants, 295–7, 317; see
    also Transplants
    banking, 299
    plus hearts, 301
Luria, Salvador, 160
Lutherans, and artificial insemina-
    tion, 84

McConnell, James, 218
McGaugh, James, 220
McKeown, Thomas, 183
Makerere Hospital, Uganda, 49–50
Man-animal hybrids, 115–16
Maple syrup urine disease, 144–5
Marijuana, 234
Marriage
    age of, 63, 67–8
    incentives and disincentives, 72–6
    'shot-gun', incidence of, 20, 66
Mather, Kenneth, 119
Maynard Smith, John, 119–20, 124,
    125
Medawar, Sir Peter, 120, 148, 323,
    324
Memory, 217–21, 226–7
Meningitis, 187
Mental retardation, 157
Mental subnormality, 212–13
Merrill, Dr John, 286
Metrazol, 220
Miscarriages, 162

Mongolism, 100, 134–5, 207
    increase of, 187–8
'Mood pills', 233–9
Moore, Dr Francis D., 303
Morris, Dr John McLean, 36
Moyer, Dr K. E., 236–7
Muller, Hermann, 122
Murray, Dr Joseph, 286
Muscular dystrophy, 132, 141, 143
Mutations, 96–7, 139–40, 150–51

National Health Service, 21, 46,
    326–7, 337, 344
National Heart Hospital, 319
Nature, 110
Neptune, W. B., 298n
New Scientist, 77
New Zealand, contraception in, 22
Nicotine, 220

Oestrogen, 29, 30–31, 36
Oral contraceptives, see Pills, con-
    traceptive
Organs, artificial, see Artificial
    organs
Organs, transplant, see Transplants

Paraplegia, 157
Parkinson's disease, 229
Peberdy, Dr Dorothy, 26
Perinatal mortality, 181
Petronovich, Lewis, 220
Phenistix test, 192
Phenylketonuria (P.K.U.), 142, 148,
    149–51, 157, 191–4, 202, 205
Pickering, Sir George, 325
Picrotoxin, 220
Pills, abortion, 49–51
Pills, contraceptive; see also Contra-
    ception
    black market in, 51
    combined type, 30
    cost of developing, 32
    effectiveness of, 30
    male, 33

    medical risks of, 28–9, 32, 51
    post-coital, 35–7
    sequential type, 30–31
    unrestricted sale of, 29
    use of, 22, 30
Pincus, Professor Gregory, 107
Pius XII, Pope, 84
Poliomyelitis, 187, 337
Population:
    attitudes to, 54–7
    birth- and death-rates, 70–71
    control methods, 58, 72–6
    control of, 52–76
    family size, 61–2, 64–6, 69–71
    forecasting, 55, 57, 59, 69
    growth rates, 59
Prednisone, 289
Prematurity, 182, 189–91
Progestogen, 30–31, 35, 36
Prostaglandins, 49–51
Pyloric stenosis, 188–9

Queen Charlotte's Hospital, Lon-
    don, 168

Recessive Genetic Defects, 128,
    130–32, 141–5
Reemtsma, Professor Keith, 302,
    323
Retinal aplasia, 129
Rhesus blood disease, 172–3
Ribonucleic acid (RNA), 155–6,
    218–20
Road-death prevention pro-
    grammes, 349–50
Rogers, Dr Stanfield, 157–8
Rowntree, Griselda, 21n
Royal Free Hospital, London, 251
Rubin, Dr Bernard, 92

Sanger, Frederick, 203
Scandinavia:
    abortion in, 41, 45, 50
    cystic fibrosis in, 142
    infant mortality in, 185
    pregnant brides percentage, 66

Schofield, Michael, 24
*Science,* 177
Scotland, anencephaly rate in, 182
Screening for disease, 164, 332–3,
  343–7
  genetic, 145–51
Scribner, Dr Belding, 226, 260, 263
Seattle, U.S.A., kidney machine
  unit in, 260
Selective breeding, *see* Eugenics
Sex:
  education, 25–6, 72
  hormones, 30–31, 36
  morality of, 25
  prevention by fear of pregnancy,
    25
  purpose of, 23–4
  teenagers and, 24–5
Sex choice, 105–10, 113–14, 144–5
  cloning as sure method of, 113–14
Sex-linked genetic defects, 132–5,
  141, 168
Sex ratio, 108–9
Shaldon, Dr Stanley, 253, 262, 263
Shannon, Dr James, 146
Shettles, Dr Landrum, 106–7, 168
Shope papilloma virus, 158
Shumway, Professor Norman, 298,
  300
Sickle-cell anaemia, 142
Simon, Lionel, 221
South Africa, heart transplants in,
  297–300
Soviet bloc:
  abortion in, 41, 47–8
  sex ratio in, 109
Sperm banks:
  A.I.D., 93–9, 123–5
  A.I.H., 83–5
Spina bifida, 196–201, 202, 207, 209
Starzl, Dr Thomas, 292
Steptoe, Patrick, 100, 101
Sterilization, 23, 33–5, 72, 96
  eugenic, 137, 141
  laws, 137–9

Stilboestrol, 36
Stillborn babies, 126, 181
'Stochastic ethics', 138
Strychnine, 220
Subfertility, 80–104
  clinics, 81–2, 83

Tatum, Dr Edward, 157
Tay-Sachs disease, 169
Taylor, Gordon Rattray, 111–12
Teratogens, 184–5
Test-tube babies, *see* Artificial
  womb
Test-tube fertilization, 99–102
Test-tube reproduction, 80–116,
  179–80; *see* Artificial Insemina-
  tion, Subfertility
Thermo Electron Corporation
  (U.S.A.), 274
Tietze, Dr Christopher, 34n
*Times, The,* 77
Tizard, Jack, 212
Townsend, Professor Peter, 333
Transplants, 284–325, 341; *see also*
  Heart, Kidney, Liver, Lung
  animal donors, 285, 291, 299,
    323–5
  'black market' in organs, 318
  dead donors, 285–6, 288, 290–91,
    298, 306–15
  death, definition of, 309–12
  early history of, 285–91
  legal position, 104n, 314
  living donors, 286, 288, 290–91,
    306–7
  organ storage, 296, 299
  organ supply and demand, 277–9,
    308–9, 315–19
  propriety of, 270–71, 302–5
  rejection and anti-rejection drugs,
    288–92
  tissue typing, 292–3
  world tissue service, 319–23
Turner's syndrome, 133
Twins, identical, 111–15

Uganda, abortion and, 50
UNESCO, 237
United States of America:
  abortion and, 39, 45–7, 50, 170
  artificial hearts in, 267, 269
  artificial insemination and, 90
  contraception in, 22, 32
  cystic fibrosis in, 142
  death-rates, 328
  disability prevention programmes
    in, 350
  doctors' visits in, 340n
  effects of heart trouble in, 277–80
  first sperm bank in, 95
  genetic engineering in, 153, 155,
    157–8
  health costs, 328, 333, 335–6
  heart implants in, 274
  I.Q.s in, 121
  life expectancy in, 330
  medical aid in, 333, 335–6
  population growth, 54, 69
  pregnant brides percentage, 66
  prematurity in, 190
  sex determination in, 106–7, 168–
    9
  test-tube reproduction in, 179–80
  transplants in, 285, 286, 294, 295,
    296, 297n, 314n, 316, 298–9
  use of kidney machine in, 253-4
  'womb-risk' lawsuits, 171–2

Waddington, Professor C. H., 138
Washkansky, Louis, 297, 300, 305
West, Frederick, 319
West Germany, sex ratio in, 109
Wheatley, Lord, 90
Woodruff, Professor, 292
World Health Organization, 19

Yolles, Dr Stanley, 233
Yugoslavia, abortion in, 41

Zamenhoff, Dr Stephen, 176
Zuckerman, Sir Solly, 54n
Zygote, 36–7, 143

# MORE ABOUT PENGUINS
# AND PELICANS

*Penguinews*, which appears every month, contains details
of all the new books issued by Penguins as they are
published. From time to time it is supplemented by
*Penguins in Print*, which is a complete list of all available
books published by Penguins. (There are well over three
thousand of these.)

A specimen copy of *Penguinews* will be sent to you free
on request, and you can become a subscriber for the price
of the postage. For a year's issues (including the complete
lists) please send 30p if you live in the United Kingdom,
or 60p if you live elsewhere. Just write to Dept EP,
Penguin Books Ltd, Harmondsworth, Middlesex,
enclosing a cheque or postal order, and your name will be
added to the mailing list.

Note: *Penguinews* and *Penguins in Print* are not available
in the U.S.A. or Canada

# DRUGS

Medical, Psychological, and Social Facts

*Peter Laurie*

Second Edition

What are the known facts about the 'dangerous' drugs? What actual harm, mental or physical, do they cause? Which of them are addictive, and how many addicts are there?

Peter Laurie has talked with doctors, policemen, addicts, and others intimately involved with this problem. He has tried some of the drugs for himself and closely studied the medical literature (including little-known reports of American research). The result of his inquiries into the pharmacological uses and social effects of drugs today appears in this book.

Originally published as a Penguin Special which went through five printings, *Drugs* was the first objective study to offer all the major medical, psychological and social facts about the subject to a public which is too often fed with alarmist and sensational reports. For this second edition in Pelicans Peter Laurie has added fresh information and statistics concerning English users of drugs and noted changes in the law.